# LOW TEMPERATURE PHYSICS

# LOW TEMPERATURE PHYSICS

BY

M. AND B. RUHEMANN

CAMBRIDGE
AT THE UNIVERSITY PRESS
1937

# CAMBRIDGE
## UNIVERSITY PRESS

University Printing House, Cambridge CB2 8BS, United Kingdom

Published in the United States of America by Cambridge University Press, New York

Cambridge University Press is part of the University of Cambridge.

It furthers the University's mission by disseminating knowledge in the pursuit of education, learning and research at the highest international levels of excellence.

www.cambridge.org
Information on this title: www.cambridge.org/9781107671515

© Cambridge University Press 1937

First published 1937
First paperback edition 2014

A catalogue record for this publication is available from the British Library

ISBN 978-1-107-67151-5 Paperback

# PREFACE

We have attempted in this book to discuss the principal problems that have occupied low temperature physicists since the time when low temperatures began to form a separate branch of experimental science. In our general arrangement we have followed the line of gradual penetration from such macroscopic phenomena as condensation and fusion to processes intimately connected with our concepts of elementary particles, such as give rise to magnetic moment and electrical conductivity. This arrangement, naturally enough, follows fairly closely the historical sequence of events, but is free from the monotony of chronological classification.

We have purposely neglected such fields of research as have been copiously treated in textbooks and monographs, as for instance supra-conductivity and the theory of specific heats. In these cases we have dealt merely with the latest developments. On the other hand, we have given particular attention to fields that have not to our knowledge as yet been dealt with in connected form, such as the subject of crystal structures stable at low temperatures. We believe that this principle justifies the very unequal length of the various parts and chapters. Moreover, we have omitted a number of investigations which, though valuable in themselves, are imperfectly connected with the general trend of low-temperature physics, such as Vegard's very interesting work on the emission spectra of solidified gases.

The rapid development of low-temperature engineering and the numerous unsolved problems that it offers have led us to lay considerable stress on the principles of gas liquefaction and rectification, more especially as most of the work in this domain is not to be found within the scope of general physical literature.

We have taken pains to avoid the rigidity and professionalism of a textbook, which we believe is alien to our subject. For low-temperature physics specialises neither in the objects of its research nor in any particular properties of these objects, but merely in its methods of approach. It has therefore never claimed the self-sufficiency of such branches as electromagnetism and thermodynamics. On the contrary, it is intimately bound up with these and all other branches of physics and has no reason to disavow these connections. Our efforts to draw nearer to absolute zero are not merely the hectic desires of a record-hunter but are dictated by a genuine curiosity as to the properties of matter, irrespective of whether they are to be measured in amperes, angström units or calories. Half a century's experience has taught us that as long as we are in a position to attain yet lower temperatures, there will always be something of interest to study there, even if it is but those processes with the help of which the temperature has been lowered. No one seriously believes that because five-thousandths of a degree is the lowest limit hitherto reached, there is no point in attempting to go farther.

Though this book may be of some use to the specialist, we have had in mind as prospective readers rather physicists specialising in other fields and more or less passively interested in low-temperature work and students who have not yet concentrated on one particular branch of physics. On the whole we have deemed it preferable to be too elementary for the former than too "advanced" for the latter.

M. AND B. R.

KHARKOV, U.S.S.R.
*December* 1935

# CONTENTS

⟨ vii ⟩

## PART IV

# THE "FREE" ELECTRON

# Part I

## PHASE EQUILIBRIUM

---

### CHAPTER I
### EARLY METHODS OF GAS LIQUEFACTION

#### I. I. 1.  *December* 1877

On 2 December 1877 Louis Cailletet, an engineer of Châtillon-sur-Seine, wrote the following letter to Sainte-Claire Deville: "Without losing a minute I hasten to inform you that I have to-day liquefied carbon-monoxide and oxygen. Maybe I am mistaken when I say 'liquefied', for at a temperature of − 29°, obtained by evaporating sulphur-dioxide, and at a pressure of 200 atmospheres, I saw no actual liquid; but under these conditions so thick a mist appeared that I concluded it was vapour near the point of liquefaction...."

On 22 December the secretary of the French Academy of Science received a telegram from Geneva: "Oxygen liquefied to-day at 320 atm. and 100 degrees cold with sulphur dioxide combined with carbon dioxide Raoult Pictet."

These events, which we now consider as marking the first step into a new domain of science, appeared at the time as the culmination of half a century's concentrated labours. When in 1823 Faraday heated chlorine hydrate at one end of a closed V-shaped test-tube, producing liquid chlorine at the other end, little or nothing was known concerning liquid vapour equilibrium. After a number of gases had been liquefied at room temperature by simple compression, scientists were at a loss to account for the failure of this method when applied to oxygen and hydrogen. As the influence of temperature was dimly realised, various investigators began cooling their gases with the slender means then available, but no positive results

RLTP ⟨ 1 ⟩ I

were obtained. The expression "permanent gases" was coined, though Faraday himself was firmly convinced that all gases could be liquefied at sufficiently low temperatures.

The immense technical possibilities offered by the liquefaction of gases, which have in recent years induced industrial concerns to invest millions in liquefaction and rectification plants, had in the first half of the nineteenth century occurred to no one. To obtain air in the liquid state was a purely scientific problem and engrossed a number of the best scientists of the time, whose work was followed with intense interest by the whole scientific world.

Falteringly, as a result of innumerable unsuccessful experiments the laws of phase equilibrium were made plain. It was in the course of investigations on liquids with much higher boiling points than air that Cagniard de la Tour lighted on critical phenomena. The disappearance of a meniscus in the middle of a closed tube when the liquid was heated suggested a limiting temperature above which no increase of pressure can lead to liquefaction in the ordinary sense of the word, i.e. to the sudden formation of a new phase. The general lines of liquid vapour equilibrium were laid down in 1863 as the result of Andrews' work. Andrews' $CO_2$ isothermals have appeared in so many textbooks that it seems unnecessary to reproduce them here. They contain almost everything that is to be said on liquid vapour equilibrium of one-component systems. The results may be summarised briefly as follows:

(1) Every one-component gas possesses a well-defined critical temperature, characterised on the corresponding $p$-$v$ isothermal by a point of inflexion, the tangent to which is parallel to the $v$-axis.

(2) The co-ordinates of the point of inflexion on the critical isothermal determine the critical pressure and critical volume of the gas.

(3) At temperatures below the critical, the gas may be liquefied by simple compression, the pressure remaining constant during liquefaction. The coexistent phases thus appear as horizontal straight lines in the corresponding isothermals, when $v$ is chosen as the axis of abscissae.

The pressures needed to liquefy a gas are not high. The critical pressures of the so-called permanent gases are less than 50 atm. and those of most liquids apart from molten metals do not exceed 100 atm.

(4) The further the temperature is lowered below the critical, the smaller the pressure needed for liquefaction.

(5) Compression along the critical isothermal leads to complete liquefaction at the critical point without change of volume.

(6) At temperatures above the critical, isothermal compression does not lead to the formation of a coexistent liquid phase. Though the density may be increased far above the critical, the substance remains homogeneous. By lowering the temperature of the compressed gas below the critical, it may be brought continuously to a state that is obviously liquid without passing through a region of two coexistent phases. There is therefore no point in differentiating between gas and liquid at temperatures and pressures above $T_c$ and $p_c$.

To avoid an error that frequently occurs it should be remarked that the critical isothermal differs in no respect from other isothermals except at the critical point. No sudden change occurs in the properties of the substance in passing through the critical isothermal at any other point.

The knowledge of these facts revealed as the first problem in the liquefaction of the "permanent" gases the determination of their critical constants. Simultaneously a second problem had to be solved: by what means can the conditions necessary for liquefaction be produced? Experiments had already made it evident that the critical temperatures of these gases are low, lower in fact than those that could at the time be produced in the laboratory. Thus the liquefaction of gases became identical with the production of low temperatures.

The events that led up to and followed Cailletet's famous letter are described very vividly by Georges Claude, himself an old pioneer, who can still recall the tense excitement roused by the discovery. While Cailletet was compressing acetylene in a glass tube with a powerful hydraulic press driving a column of mercury, a valve was accidentally left open. The pressure was thus suddenly relaxed; the expanding gas drove the mercury down the tube and was cooled so far as to show a thick mist of condensing liquid. Cailletet, who was repeating Andrews' experiments, was quick to perceive what had occurred. He repeated the experiment with oxygen, cooled previously with evaporating sulphur dioxide, and the result is contained in his letter to Sainte-Claire Deville. In his book, *Air Liquide, Oxygène, Azote,* Claude describes how Cailletet refrained from making his discovery public at the next meetings of the French Academy. He had been proposed as a corresponding member and did not wish to bias the voters by a sensational publication before the election. On 24 December the announcement of his discovery was to take place. Two days earlier the Academy received Pictet's telegram, stating that he had liquefied oxygen. The letter to Sainte-Claire Deville is the only proof of Cailletet's priority.

Cailletet's experiment can hardly be said to have solved the problem of air liquefaction. Under his conditions a few drops of liquid could at the best be produced, which must needs evaporate the moment the expansion ceased. However, in view of the fact that this method has of late in a modified form become of importance for the liquefaction of helium, it may be well to show how a simple calculation may roughly account for Cailletet's result. Suppose that we may as a first approximation neglect the transmission of heat from the gas to the mercury and to the glass walls of the tube, an assumption that might be justified if the expansion were sufficiently rapid, then our case is a simple adiabatic expansion, for which the First Law gives the differential equation

$$du = -p\,dv, \quad q \text{ being zero.}$$

Assuming the gas to obey Boyle's law under the conditions of the experiment, an hypothesis which is of course far from correct, and taking $v$ and $T$ as independent variables, we have

$$C_v dT = -RT/v\,dv.$$

Separating $v$ and $T$, we may integrate the equation and obtain

$$\log T_1/T_2 = R/C_v \cdot \log v_2/v_1 = R/C_p \cdot \log p_1/p_2.$$

The molal heat of oxygen at constant pressure is approximately $7R/2 \sim 7$ cal./$^\circ$; $p_1$ in Cailletet's case was in fact 300 atm. and $T_1$ about 250° K. Inserting these figures, we obtain

$$\log 250/T_2 = 2/7 \log 300 \sim 0\cdot 7,$$

which gives    $250/T_2 \sim 8\cdot 5$    or    $T_2 \sim 29°$ K.

We now know that the normal boiling point of oxygen is 90° K. It is thus plainly within the bounds of possibility to produce liquid oxygen by Cailletet's method. But as the latent heat of evaporation of oxygen is 1600 cal./gramme-molecule, it is not surprising that only a few drops are obtained.

We have already stated that Pictet very nearly forestalled Cailletet in liquefying oxygen. His work is important not only in exemplifying the laws of phase equilibrium but because the method he first employed, and which was later perfected by Kamerlingh Onnes in Leiden, has proved a most powerful instrument for air liquefaction in the laboratory.

In Pictet's so-called Cascade Process the low temperature needed for the liquefaction of oxygen is obtained gradually in the course of several stages. In the first a gas easily liquefied at room temperature is condensed under pressure and expanded into a separate low-pressure chamber, in which the vapour is pumped back to the compressor, the gas thus circulating through the system. Through evaporation the temperature in the low-pressure chamber can be reduced below the critical point of a second gas. In the second stage this second gas can then be condensed and thereupon cooled in a second low-pressure chamber by evaporation. The process may be continued until the critical temperature of oxygen is reached; and in this way air, nitrogen and carbon monoxide may be liquefied.

Pictet in his first experiments used $SO_2$ in his first circuit and $CO_2$ in the second. In the low-pressure chamber of the $CO_2$ circuit the liquid carbon dioxide was solidified and cooled by sublimation. Oxygen at high pressure was admitted to a tube immersed in the solid $CO_2$ and then released through a valve to the atmosphere. Pictet describes how a jet of liquid oxygen shot through the valve and immediately evaporated. However, though he himself believed that a temperature of $-140°$ C. in the $CO_2$ was reached, it is almost certain that it remained above $-110°$ C. As the critical temperature of oxygen is now known to be $-118°$, we must conclude that the liquid oxygen was formed on releasing the pressure as in Cailletet's experiment.

In Kamerlingh Onnes' cascade at Leiden, which was working until quite recently and with the help of which all the notable cryogenic experiments were carried out, the first circuit contained methyl chloride, the second ethylene and the third oxygen.

### I. 1. 2. *The Cracow School*

The first experiments of Cailletet and Pictet were followed almost immediately by the installation at Cracow in Poland, then in the Austrian province Galicia, of a cryogenic laboratory in the most advanced sense of the word. Here Wroblewski and Olszewski, equipped with highly efficient scientific apparatus, proceeded to "create" Low Temperature Physics. From a glance at any of Olszewski's papers of the early eighties the purpose of these experiments is immediately evident. A few drops of fog in a tube and an evaporating jet of liquid are no basis for scientific research. What we need is a liquid "boiling quietly in a test-tube", which we can observe at leisure, and the characteristic properties of which we can determine accurately. By the production of condensed gases an entirely new group of bodies has entered our field of vision. Our next object is plainly to find out all that we can about them. Ten years after the discovery of liquid oxygen, Olszewski had liquefied several cubic centimetres of all gases known at the time except hydrogen, had determined their boiling points with considerable accuracy as well as their critical temperatures and pressures, had solidified all except oxygen and measured their triple points, had measured the densities of the liquefied gases boiling under atmospheric pressure and had even observed the absorption spectrum of liquid oxygen. It is true that none of Olszewski's determinations has survived the intervening half-century, but in all cases a repetition of his experiments under the more congenial conditions of to-day has led only to a slight correction as a result of greater accuracy in the measurement of temperature and pressure and greater purity of the material investigated. In 1887 Olszewski knew almost as much about liquefied gases as we know to-day. When we remember that the era of commercial gases had not yet begun, that there were no Dewar vessels and no high vacuum pumps, we can but admire the work of the Polish scientists, carried out in a provincial university far away from the centres of learning of their time.

A short survey of Olszewski's early apparatus may serve as an introduction to low-temperature experimental technique and to the more highly developed appurtenances of the Leiden Laboratory. One of the most successful innovations of the Polish school was the introduction of liquid ethylene as a cooling agent. The advantage of this substance is that it may be liquefied slightly below room temperature at moderate pressure, whereas a vapour pressure of 10 mm. of mercury corresponds to a temperature of $-150°$ C. This temperature is well below the critical temperatures of oxygen, carbon monoxide and nitrogen, so that all these gases may be liquefied in liquid ethylene by simple compression, without employing Cailletet's expansion principle. However, in order to study them at atmospheric pressure and below, at still lower temperatures, it was necessary to devise some method of isolating them from the surrounding ethylene cryostat. This was accomplished with considerable success, as shown in fig. 1. Olszewski kept his liquid ethylene in a steel flask containing a syphon and immersed in a bath of ice and salt. Under these conditions the vapour pressure of ethylene is 32 atm. The liquid then passed through a copper spiral contained in a bath of $CO_2$ and ether, the vapour pressure of which was reduced by a vacuum pump so as to give a temperature of about $-100°$ C. Through this spiral the ethylene was expanded to atmospheric pressure into a glass vessel $g$. Part of the ethylene naturally evaporated on expansion, but by keeping the $CO_2$ vessel sufficiently cold, 230 grm. of ethylene could be made to give 150 c.c. of liquid in a glass vessel at atmospheric pressure, i.e. at a temperature of $-104°$ C. This was sufficient for a number of experiments.

Inside the glass vessel $g$ were placed a number of concentric glass tubes $t$ closed at the bottom, into the innermost of which the liquid ethylene was introduced. Thus, when the vapour pressures were further lowered by pumping through $n$, the cold ethylene vapour passing round the intervening glass tubes served to shield the liquid from the warm air of the atmosphere. This protection, together with that offered by the concentric glass walls, proved sufficient to keep the liquid ethylene at 10 mm. pressure for a considerable time.

Olszewski's actual experiments with liquid gases were carried out inside a thick-walled glass tube $a$, immersed in the liquid ethylene and sealed to a brass cap $c$ traversed by two narrow metal tubes. One of these was connected to a hydrogen gas-thermometer vessel $h$ inside $a$, the other to a metal flask containing at high pressure the gas to be liquefied. To the glass tube $a$, which was 30 cm. long and had an inner diameter of 14 mm. and walls 3·5 mm. thick, Olszewski devoted great attention. By very carefully annealing the tube he was able to apply pressures of 60 atm. for long periods as well as sudden changes of temperature. In order to study the condensed gases at atmospheric pressure and below, the tube $a$ contained yet another tube $u$ closed at the bottom. The gas first condensed in the outer

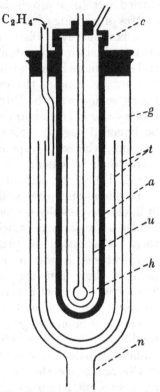

Fig. 1. Olszewski's apparatus for experiments with liquefied gases.

annular space between $a$ and $u$, then rose to the top of $u$ and flowed over. When the vapour pressure was reduced the liquid naturally evaporated first from the annular space, which through the glass wall of $a$ was in contact with the liquid ethylene. The liquid in the innermost tube was thus shielded from the liquid ethylene by a cylinder of gas at low pressure as well as two glass walls. The insulation, which appears primitive enough to-day, enabled Olszewski to pump off solid nitrogen well below the triple point, reaching a minimum temperature of $-225°$ C., as read on his hydrogen thermometer. A careful study of Olszewski's papers will show that the accuracy of his temperature determinations in this region was probably better than $\pm 2°$. $-224°$ C. is still the lowest temperature that we can hope to reach with solid nitrogen without employing very complicated methods.

One of Olszewski's first experiments was to combine his refrigerating system with an expansion device according to Cailletet for the purpose of liquefying hydrogen. A narrow glass tube filled with gaseous hydrogen, capable of withstanding 200 atm., was bent twice at right angles. One end was immersed in liquid nitrogen, the other connected to a mercury pump as in Cailletet's experiments. When the mercury pressure was suddenly released, a fog of liquid hydrogen was observed in the tube, which was immediately obliterated by a fog of solid nitrogen. But Olszewski was unable by this method to obtain liquid hydrogen "boiling quietly in a test-tube". The liquid hydrogen evaporated as soon as it was formed, and its production in sufficient quantities to enable measurements to be made was the outcome of a new principle which will be discussed in the next chapter.

It may appear unfair to Olszewski, whose experiments were continued with great success until well into the present century, to break off at this point our description of his work. Yet towards the end of the nineties the Crakow school fell into the shadow of new developments originating in other countries, and before the close of the century the centre of gravity of low-temperature research had shifted definitely to Holland.

### I. I. 3. *Van der Waals' Equation and the Law of Corresponding States*

With the help of concepts developed in the kinetic theory of gases van der Waals in 1873 introduced his well-known equation of state. Assuming the molecules of a gas to occupy a finite volume and to exercise well-defined forces on one another, he thus gave the first hypothesis concerning the laws of interaction of material particles. However summary and incomplete this picture of a molecule must necessarily be, van der Waals' equation has been extraordinarily successful in describing qualitatively and to a certain extent quantitatively the properties of real and condensed gases. Half a century elapsed before quantum mechanics in the hands of London gave a more detailed and comprehensive account of the nature of van der Waals' forces.

Writing van der Waals' equation in the form

$$(p + a/v^2)(v - b) = RT,$$

we shall develop as briefly as possible the consequences to which it leads us.

The equation being cubic in $v$, we see at once that the isothermal curves will in general have a maximum and a minimum, which may or may not be real. It is hardly necessary to show how these maxima and minima may be superposed on Andrews' isothermals, giving the unstable and metastable homogeneous states, which may exist apart from the stable two-phase condition. The laws of thermodynamics show that Andrews' horizontal straight lines, marking the states of liquid vapour equilibrium, must be so situated that the closed areas cut off in van der Waals' curves above and below these lines are equal to one another. The equation, which contains the three constants $a$, $b$ and $R$, enables us to compute the critical temperature, pressure and volume as a function of these constants. Indeed the critical point, which we know to be given by a horizontal point of inflexion, can be expressed mathematically by setting the first two derivatives of $p$ by $v$ equal to zero. This leads to two new equations:

$$\left(\frac{\partial p}{\partial v}\right)_T = -\frac{RT_c}{(v_c - b)^2} + \frac{2a}{v_c^3} = 0 \quad \text{and} \quad \left(\frac{\partial^2 p}{\partial v^2}\right)_T = \frac{2RT_c}{(v_c - b)^3} - \frac{6a}{v_c^4} = 0,$$

which, together with the equation of state itself, are sufficient to compute $T_c$, $v_c$ and $p_c$ as functions of $a$, $b$ and $R$. The result is

$$T_c = 8a/27bR, \quad v_c = 3b, \quad p_c = a/27b^2.$$

Similarly, $R$, $a$ and $b$ may be derived from the critical constants of the gas. This is the foundation of van der Waals' Law of Corresponding States, which may be derived as follows:

Let $T/T_c = T'$, $v/v_c = v'$, $p/p_c = p'$.

We may then insert $T_c T'$ for $T$, $v_c v'$ for $v$ and $p_c p'$ for $p$ in the equation to state and obtain

$$\left(\frac{p'a}{27b^2} + \frac{a}{v'^2 9b^2}\right)(3v'b - b) = \frac{T'8a}{27b},$$

whence, cancelling $a/9b$, we have

$$\left(\frac{p'}{3} + \frac{1}{v'^2}\right)(3v' - 1) = \tfrac{8}{3}T',$$

an equation which should be uniformly valid for all gases and contain only implicitly the critical constants.

Now the constants $R$, $a$ and $b$ for a given gas may be determined from isothermals measured at ordinary temperatures, and we see that with the help of these constants the critical data may be computed. Van der Waals' equation therefore enables us to determine from measurements made at higher temperatures the conditions necessary for the liquefaction of a gas, even when the critical temperature of the gas is considerably lower than the temperatures at which the measurements have been carried out.

The Law of Corresponding States is not accurately valid any more than van der Waals' equation of state. But it is an ex-extremely useful guide, and the concept which it expresses is fundamental. For without it we should be at a loss for a clear definition of what is meant by low temperatures. Since every substance is characterised by its critical constants $T_c$, $v_c$ and $p_c$, and may, by employing the reduced variables $T'$, $v'$ and $p'$, be brought to coincidence with any other substance, it is clear that a given temperature, which may be considered low for one substance, will appear high for another. We may, for instance, easily liquefy $CO_2$ at room temperature by applying a pressure of some 30 atm., whereas a much lower temperature is required for the liquefaction of oxygen. It is therefore reasonable to call room temperature high for air and comparatively low for $CO_2$, whereas it should naturally be considered extremely low for water and even lower for platinum. We shall find it profitable to call a given temperature high or low for a given substance according as it is high or low compared with the critical temperature of the substance. A great number of differences in the behaviour of various substances at the same temperature will thus become clear from the point of view of corresponding states.

# CHAPTER II

# INDUSTRIAL AIR LIQUEFACTION

## I. II. 1. *General Principles*

Neither Cailletet's nor Pictet's method was destined to play an important part in large-scale gas liquefaction. Let us shortly consider the problem from the point of view of engineering and commerce. The units of the engineer are kilowatt hours, those of commerce are pounds, shillings and pence. Industry has combined these sets of units with the help of certain equations: so much work costs so much money. The problem is to produce as much liquid air with as little work as possible, i.e. as cheaply as possible.

The fact that work is necessary at all to liquefy air is due to the second law of thermodynamics and is connected with the fact that to remove heat from a body its entropy must be decreased. This entails the expenditure of work. A machine which accomplishes this is called a refrigerator.

A gas liquefier as conceived by the engineer is a refrigerating machine in the strict sense of the word. A certain amount of energy must be withdrawn at low temperature, conveyed to room temperature and there given off to the surroundings. Moreover, the temperature drop between room temperature and that of liquid air must be brought about and maintained throughout the process.

The amount of work required in theory by a refrigerating machine is given by the laws of thermodynamics and may best be shown with the help of the entropy diagram. Fig. 1 shows a schematical entropy diagram of air. The ordinates are temperatures, the abscissae entropy. The sloping lines represent isobars, the curve the boundary line between the liquid and gaseous states. The figure shows that the entropy of liquid air, $S_0$ (point $D$), is smaller than that of gaseous air at room temperature and atmospheric pressure, $S_1$ (point $B$). The entropy $S_1 - S_0$ must be given off at room temperature. This is usually effected by transferring heat to the water of a compressor. We shall suppose this to take place isothermally at room temperature $T_1$. Then the quantity of heat to be transferred is given by the equation

$$Q_1 = T_1 (S_1 - S_0).$$

On the figure this is equivalent to the area of the rectangle $ABEF$. The heat $Q_2$ which must be removed from the gas in order to liquefy it consists of the heat of vaporization $\lambda$, which

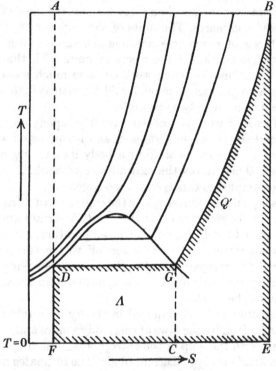

Fig. 1. Schematical entropy diagram of air.

must be absorbed at a constant low temperature $T_0$, and the heat $Q'$ required to cool the air from room temperature to the boiling point. On the figure this is represented by the area $BGDFE$. Now if we consider the water, which takes up the heat $Q_1$, and the gas as a single system, the minimum work $A$ required is given by the condition that the entropy increase of the water shall be equal to the entropy decrease of the gas, i.e.

$$S_0 - S_1 + Q_1/T = 0,$$

or since $Q_1 = Q_2 + A$,

$$A = T_1(S_1 - S_0) - Q_2 = T_1(S_1 - S_0) - \lambda - Q',$$

and is thus equal to the area $ABGD$.

In the case of air the actual figures are

$$S_1 = 0 \cdot 90, \ S_0 = 0, \ Q_2 = 97 \cdot 1 \ \text{k.cal.}$$

Thus for $T_1 = 290$ we obtain $A = 160 \cdot 7$ k.cal. or

$$A = 0 \cdot 254 \ \text{H.P.h.} = 0 \cdot 19 \ \text{kw.h. for 1 kg.}$$

The above computation was made on the assumption that only the heat of vaporisation is removed at the boiling point of air. In the Carnot cycle, which considers the case of two heat reservoirs at different constant temperatures $T_0$ and $T_1$, heat being transferred from the colder to the warmer reservoir, the entire heat absorption takes place at the lowest temperature $T_0$. To remove a quantity of heat $Q_2$ from the low-temperature reservoir, a quantity of work $A'$ is required, which is given by the well-known formula

$$A' = Q_2 \frac{T_1 - T_0}{T_0}.$$

If we put $Q_2 = 97 \cdot 1$ k.cal., $T_1 = 290°$ K., $T_0 = 81°$ K., then

$$A' = 0 \cdot 40 \ \text{H.P.h.} = 0 \cdot 30 \ \text{kw.h.}$$

The Carnot process is thus less economical than the "ideal" process, in which part of the heat is absorbed at higher temperatures, according to the well-known principle that the lower the temperature of a body with respect to its surroundings, the more work is needed to cool it yet further.

It is one thing to compute theoretically the amount of work necessary to liquefy a quantity of air and another to find a machine which will realise this process. The two processes we have sketched would require machines working entirely without loss of entropy. Consider a kilogram of air at normal pressure and temperature in contact with its surroundings.

To convert this into liquid air, entropy must be removed. How can this be effected? If entropy were a function of temperature alone, it would be impossible; for the low temperature, with the help of which the air could be cooled, has yet to be attained. Fortunately the entropy depends on volume as well as on temperature, and the volume may be changed with the help of work. As we see from fig. 1, and as we can easily deduce from the Second Law, entropy decreases with increasing pressure, i.e. with decreasing volume. Thus if we expand a gas adiabatically, thus maintaining its entropy constant, we may decrease its temperature. On the other hand, by compressing a gas isothermally we may decrease its entropy, this entropy being absorbed by the surrounding medium. The ideal liquefier would thus have to work as follows: we must compress the air isothermally to point $A$ (fig. 1) and then expand it adiabatically to point $D$, thus liquefying the entire gas. But the diagram would show us that the pressure would have to be immense. There appears to be no possibility of realising the "ideal" process. Not even an efficiency corresponding to the Carnot cycle can in fact be reached, as we shall shortly demonstrate. We must therefore be content with machines working with loss of entropy and attempt to keep these losses as small as possible.

From the above it is clear that a liquefier must necessarily consist of three parts, a cold part, in which the gas is liquefied, a warm part, in which the heat removed from the gas is given off to the surroundings, and an intermediate part connecting the other two. As in the course of the process the gas must flow from the warm to the cold part of the apparatus, the intermediate portion must be so constructed that the gas may here be cooled from room temperature to the boiling point with as little loss as possible. In all industrial liquefiers the work done by the machine consists in compressing the gas in the warm part of the apparatus. Thus "charged", the gas passes through the intermediate portion, in which it is cooled to a low temperature, and is thereupon "discharged", i.e. expanded in the cold part. Hereby a quantity of heat is absorbed, leading to partial liquefaction of the gas, the rest passing back through the intermediate portion. The fact that in all cases only part of the gas is liquefied is essential for the construction of the intermediate portion, as it enables the in-going warm gas to be cooled with the help of the out-going cold gas. This is the general principle of the heat exchanger.

Linde's original heat exchanger, which may still be inspected in the factory near Munich, consisted of two iron tubes 100 m. long and 10 and 4 cm. respectively in diameter. The narrow tube was thrust through the wider and the two were then wound to a cylindrical spiral of 32 windings, mounted in wooden scaffolding and insulated with wool. The general principle has remained up to the present day. Copper is now used instead of iron and a number of inner tubes are inserted in the outer pipe to increase the heat transfer and avoid a pressure drop between one end of the exchanger and the other. Hampson's heat exchanger consists of a number of parallel tubes wound so as almost completely to fill an outer cylinder. The compressed air is admitted through the windings, the out-going cold gas passing up between them.

Two types of apparatus have been developed for the large-scale liquefaction of air, one by Linde and Hampson, the other by Claude and Heylandt. The methods are in many points similar, in fact Heylandt's liquefier is a combination of the elements of Claude and Linde.

Claude, by employing an adiabatic expansion cylinder, lowers the temperature of the gas at almost constant entropy, letting the compressed gas do work against the piston, which is so coupled as to "help" the compressor. Hampson and Linde cool their gas with the help of the Joule-Thomson effect, which is the temperature drop undergone by a gas expanded through a throttle valve, and is due to the work done against van der Waals' forces. In both cases the temperature difference between the ends of the apparatus is maintained by means of heat exchangers.

We shall discuss the methods in greater detail in § 3 and shall here merely note that Claude's method, which is in the main isentropic, should theoretically be more effective than the Linde process, which is essentially anisentropic. That in point of fact the two processes are of fairly equal efficiency is to be explained partly by several serious technical difficulties of the

Claude principle, connected with driving an expansion machine at very low temperatures, and partly by the untiring labour with which Linde continued to improve his apparatus.

## I. II. 2. *The Joule-Thomson Effect*

In 1862 Joule and Thomson discovered that a gas, on being slowly expanded through a porous plug from a pressure $p_1$ to a pressure $p_2$, without doing external work and without gaining kinetic energy, undergoes a slight change of temperature, which varies in magnitude and sign according to the nature of the gas. Carrying out the experiment at room temperature, they found that air, oxygen and nitrogen are slightly cooled by expansion, whereas hydrogen is heated. This effect, which for a perfect gas may be shown to vanish, can be computed thermodynamically as follows.

If the molecular volumes corresponding to $p_1$ and $p_2$ be designated as $v_1$ and $v_2$ respectively, the work done against the internal forces is $p_2 v_2 - p_1 v_1$. Now according to the First Law, $Q$ being zero,

$$U_1 - U_2 = p_2 v_2 - p_1 v_1,$$

$U_1$ and $U_2$ being respectively the internal energy in the states denoted by 1 and 2. We thus see that in the course of a Joule-Thomson expansion the expression $i = U + pv$ remains constant. $i$, which is known as the enthalpy or total heat, will play an important part in the following paragraphs.

We may introduce $p$ and $T$ as independent variables by putting $di = dU + p\,dv + v\,dp = 0$. Therefore

$$\left(\frac{\partial U}{\partial T}\right)_p dT + \left(\frac{\partial U}{\partial p}\right)_T dp + p\left(\frac{\partial v}{\partial T}\right)_p dT + p\left(\frac{\partial v}{\partial p}\right)_T dp + v\,dp = 0.$$

Now according to a well-known thermodynamical relation

$$\left(\frac{\partial U}{\partial p}\right)_T = -T\left(\frac{\partial v}{\partial T}\right)_p - p\left(\frac{\partial v}{\partial p}\right)_T.$$

Therefore

$$\left(\frac{\partial U}{\partial T}\right)_p dT + p\left(\frac{\partial v}{\partial T}\right)_p dT - T\left(\frac{\partial v}{\partial T}\right)_p dp + v\,dp = 0.$$

But
$$\left(\frac{\partial U}{\partial T}\right)_p + p\left(\frac{\partial v}{\partial T}\right)_p = C_p.$$

Therefore
$$C_p dT = \left\{T\left(\frac{\partial v}{\partial T}\right)_p - v\right\} dp.$$

So we may write as the differential Joule-Thomson effect

$$\left(\frac{\partial T}{\partial p}\right)_i = \frac{T\left(\frac{\partial v}{\partial T}\right)_p - v}{C_p}. \qquad \ldots\ldots(1)$$

In the case of a perfect gas, when $v = \dfrac{RT}{p}$, we have

$$T\left(\frac{\partial v}{\partial T}\right)_p = \frac{RT}{p} = v,$$

so that the Joule-Thomson effect is zero.

From (1) it is evident that the equation

$$T\left(\frac{\partial v}{\partial T}\right)_p = v \qquad \ldots\ldots(2)$$

gives a curve along which $\left(\dfrac{\partial T}{\partial p}\right)_i = 0$. By employing as co-ordinates the reduced pressure and temperature and making use of the Law of Corresponding States, either as expressed in van der Waals' or in some other equation of state, we obtain a curve as shown in fig. 2, the so-called inversion curve of the Joule-Thomson effect. To the right of this curve the gas is warmed by expansion, to the left it is cooled. A similar curve may be obtained for the integral effect $\int_{p_1}^{p_2}\left(\dfrac{\partial T}{\partial p}\right)_i dp$. The curve obtained with the help of the van der Waals equation is not quantitatively corroborated by experiment as the equation is not accurate enough. The curve on fig. 2 was therefore not determined from this equation but taken from Meissner's more precise but more empirical data. The inversion curve shows us at once what conditions of temperature and pressure must be fulfilled so that a gas may be liquefied with the help of the Joule-Thomson effect. To cool a gas by isenthalpic expansion we must begin at temperatures and pressures corresponding to

2-2

points to the left of the curve. We can thus cool a gas from room temperature only if pressures exist for which this temperature gives such points. Now the inversion curve intersects the axis of reduced temperatures at $T' = 5.8$. To obtain a

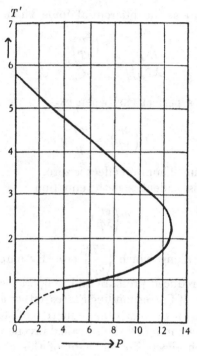

Fig. 2. Inversion curve of Joule-Thomson effect.

Joule-Thomson effect of the right sign at room temperature the reduced temperature of the gas at say 290° K. must be less than this figure. Table I illustrates the position in the case of oxygen, air, hydrogen and helium.

Hence it is immediately clear that, whereas oxygen and air may be liquefied with the Linde method, starting from room temperature, hydrogen and helium may not. To liquefy hydrogen the gas must first be cooled with liquid air, whereas helium must be precooled with liquid hydrogen in order to give the Joule-Thomson effect the correct sign. This is a great difficulty in the liquefaction of hydrogen and helium, and it is important that in recent years methods have been developed which obviate this difficulty. We shall return to this in chap. III, § 1.

Table I

| Gas | $T_c$ | $T' = 290/T_c$ |
|-----|-------|----------------|
| $O_2$ | 154·0 | 1·88 |
| Air | 133·0 | 2·19 |
| $H_2$ | 33·2 | 8·7 |
| He | 5·2 | 56 |

It is clear how the Joule-Thomson effect may be used to lower the temperature of a gas to the point of liquefaction. But when once this point is reached the temperature at the expansion valve remains constant and the Joule-Thomson effect no longer governs the process. We are henceforward concerned with the fraction of the gas passing through the expansion valve, which is actually liquefied. The liquefaction coefficient $\epsilon$ is evidently determined by the quantity of heat that is absorbed by a gas when the pressure is diminished at constant temperature. This is given by the differential expression $\left(\dfrac{\partial i}{\partial p}\right)_T$ or by its integral over a finite pressure. The former is known as the isothermal expansion coefficient. By a simple calculation Meissner showed that $\epsilon$ is a maximum for values of $p$ and $T$ outside the liquefier that lie on the inversion curve of the differential Joule-Thomson effect. This gives us the optimum condition for working a Linde liquefier. Since the temperature outside a liquefier cannot always be chosen arbitrarily—for simple air liquefiers we have room temperature, for hydrogen liquefiers the temperature of liquid air—we should attempt to choose the pressure so that the corresponding point on the $p$-$T$ diagram lies on the inversion curve. Unfortunately, in the case of air this is hard to realise. If we plot the point on fig. 2 we obtain a pressure of about 460 atm. To work at this pressure is not economical for purely secondary reasons; moreover, it can be shown that the loss sustained by lowering the pressure is not very great. Most plants are operated at about 200 atm., the practical advantages of the lower pressure outweighing the theoretical losses.

Professor Linde describes in his memoirs the critical occasion when his first apparatus was tested on 29 May 1895. On account of the low pressure (65 atm.) and the high heat capacity of the apparatus it took three days to bring the expansion valve down to the boiling point of air, part of the temperature drop obtained in the day's run being lost again during the night. Finally, on the third day "with joyous excitement we watched the temperature fall according to the law laid down by Thomson and Joule, even after the limits in which these scientists had worked had been considerably surpassed. After the stationary state had been reached, which had been determined by the physicists as corresponding to saturation or liquefaction, we continued work for a sufficient time to justify the expectation that a considerable amount of air had been liquefied. Then, amid rising clouds, we allowed the beautiful, bluish liquid to stream out into a large pail...."

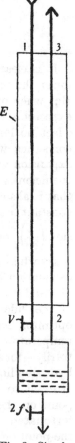

### I. II. 3. *The Efficiency of Liquefiers*

From the above it is clear that a liquefier of the Linde type may be represented schematically as in fig. 3. Compressed gas enters the heat exchanger $E$ at 1 and is expanded in the valve $V$; whereupon part is liquefied and may be tapped off at $2f$, the rest returning through $E$ and emerging at 3.

Let us now return to the engineer's point of view and ascertain how far this machine satisfies the requirements for producing liquid air profitably. For calculations of this type it is customary to employ a diagram in which enthalpy is plotted as a function of temperature at various pressures. This $i$-$T$ diagram has been determined for air and a number of other gases with great

Fig. 3. Simple Linde liquefier.

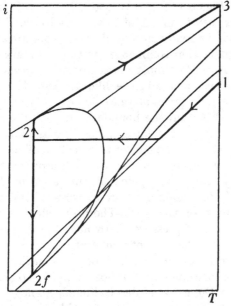

Fig. 4. Simple Linde process on $i$-$T$ diagram.

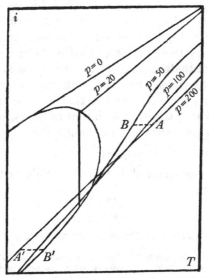

Fig. 5. Schematical $i$-$T$ diagram of air.

accuracy.* Fig. 5 shows schematically the $i$-$T$ diagram for air, which is easy to interpret in analogy to the $T$-$S$ diagram which we have already employed. If we cool air at constant pressure liquefaction commences at the boundary curve and the isobar proceeds almost vertically downwards; not quite vertically, since in a mixture like air the dew point and boiling point for a given pressure do not completely coincide. (See chap. v.) Within the boundary curve the air separates into liquid and gas, the latter possessing an enthalpy corresponding to the upper point of intersection of the isobar and the boundary curve, whereas the enthalpy of the former is given by the lower point of intersection. It is clear that, outside the boundary curve, we are here within the inversion curve of the Joule-Thomson effect; for an isenthalpic decrease of pressure (horizontal line in the figure) is connected with a decrease of temperature.

The simple Linde process is shown schematically in fig. 4, the numbers corresponding to those in fig. 3. To ascertain the efficiency, i.e. the work needed to liquefy a kilogram of air in relation to the figures found for the "ideal" and the Carnot process, it will be well to neglect the work required to cool down the apparatus. For once this is achieved a large plant can run for several weeks on end without warming up. So we will consider a liquefier in its stationary state, which is characterised by the fact that, in all parts of the apparatus, temperature, enthalpy and quantity of air present are constant with regard to time. Thus the enthalpy entering as compressed gas at 1 must be equal to that leaving in the form of expanded gas at 3, together with that of the liquid air tapped off at $2f$. This gives the continuity equation

$$i_1 = \epsilon i_{2f} + (1 - \epsilon) i_3,$$

where $\epsilon$ is the coefficient of liquefaction. Solving this equation for $\epsilon$ we obtain

$$\epsilon = \frac{i_3 - i_1}{i_3 - i_{2f}}. \qquad \qquad \dots\dots(1)$$

Now the denominator of this expression is equal to the quantity

* The $T$-$S$ diagram with isenthalpic lines can be used just as well but is less instructive, since here only energy relations are used and thus the entropy itself does not enter into the calculations.

of heat $Q$ which must be removed from one kilogram of gas to liquefy it starting from room temperature and atmospheric pressure and is thus quite independent of the conditions of the experiment. The only variable in the equation is $i_1$, which depends on the pressure of the compressed air. This, as we have already mentioned, is usually taken as 200 atm. We see that $\epsilon$ is independent of the magnitude of the Joule-Thomson effect generated in the valve in the stationary state.

Taking the pressure at 1 as 200 atm. we may determine the values of $i$ at the various points from the diagram and find $\epsilon = 0.098$. Thus only about one-tenth of the air is liquefied in $V$, the rest returning through the heat exchanger, and so in order to liquefy a kilogram of air we must send a quantity equal to $1/\epsilon$ through the liquefier. This figure enables us to determine the work needed to liquefy a kilogram. This is the work necessary to compress $1/0.098$ kg. to 200 atm. Here we shall assume the compression to be accomplished isothermally and the gas to obey Boyle's law. Then the quantity of work

$$A = \frac{RT}{\epsilon} \ln \frac{p_1}{p_3} = \underline{1.26 \text{ kw.h.}}$$

If we compare this figure with the theoretical amount of work needed, we see that the efficiency of this type of apparatus is very low. The work required is nearly five times as great as even that demanded by the Carnot process. Moreover, we must add that in actual fact $A$ must be multiplied by a factor of about 2 to account for non-isothermal compression, frictional losses in the compressor and heat losses in the liquefier. So the next question to be solved is—how can this value of $A$ be diminished?

Linde succeeded in raising the efficiency of his liquefier by two devices, one of which leads to a direct decrease in the work of compression, the other to an increase in the liquefaction coefficient $\epsilon$. The first improvement is known as the *High Pressure Circuit*. The work of compression is proportional to $\log p_1/p_3$, whereas the Joule-Thomson effect is roughly proportional to the pressure difference. Thus the work may be considerably reduced by expanding only a small portion of the gas to atmospheric pressure, about as much as is actually liquefied. The rest is expanded to an intermediate pressure of 40–50 atm. and returned to the compressor.

We obtain a type of apparatus as shown in fig. 6. Fig. 7 represents the process on the $i$-$T$ diagram. Compressed air enters the heat exchanger at 1 at a pressure $p_1$ and is expanded to an intermediate pressure $p_2$ in the valve $V_1$. One part $(1-M)$ returns to the compressor at 3; the rest, $M$, is expanded in $V_2$ to atmospheric pressure and partially tapped off as liquid at $4f$, the remaining gas passing back through the exchanger and emerging at 5. To determine $\epsilon$, consider the apparatus as a whole. In the stationary state we must have

$$i_1 = (1-M)\,i_3 + (M-\epsilon)\,i_5 + \epsilon i_{4f},$$

which gives $\quad \epsilon = \dfrac{i_3 - i_1 + M\,(i_5 - i_3)}{i_5 - i_{4f}}.$

For various values of $M$ we thus obtain different liquefaction coefficients $\epsilon$, and the greater $M$, the more will be liquefied.

The work of compression will now be given by $\quad A = RT\,(\ln . p_1/p_2 + M \ln . p_2/p_5),$ which is the *smaller* the smaller $M$, and this must again be multiplied by $1/\epsilon$ to obtain the work needed to liquefy one kilogram of air.

For a given high pressure $p_1$ we may now plot this work as a function of the intermediate pressure $p_2$. For every value of $M$ we now obtain a curve with a minimum. A set of these curves is represented in fig. 8 for $p_1 = 200$ atm. and shows first that the optimum value of $p_2$ is about 50 atm. and secondly that the process appears to become more economical the smaller $M$ is chosen. This is obviously

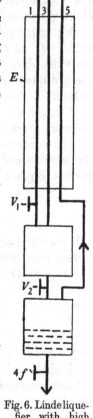

Fig. 6. Linde liquefier with high pressure circuit.

erroneous, as for $M = 0$ no liquid air would be obtained at all, and we must naturally expand as much air as is liquefied in the intermediate container, at any rate if $p_2$ is lower than the critical pressure. The question is how to find the genuine optimum of $M$.

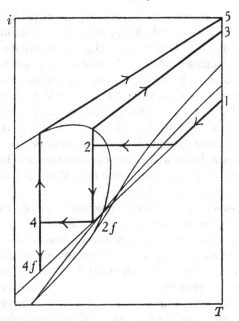

Fig. 7. Linde process with high pressure circuit.

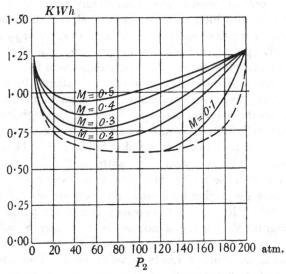

Fig. 8. Power output in Linde process with high pressure circuit.

Why does this optimum not appear in our calculation? We have hitherto considered the liquefier as a whole and paid no attention to what happens in the intermediate container behind $V_1$. In this container air may or may not be liquefied according to the adjustment of the apparatus. For instance, if $p_2$ be chosen greater than 39 atm., which is the critical pressure of air, no two-phase equilibrium can come about. Moreover, if $M$ is kept above a certain value, it is clear that the intermediate container will be drained continuously. Nevertheless, a definite constant temperature will here subsist, which we may compute with the help of our $i$-$T$ diagram for given values of $p_1$, $p_2$ and $M$.

Now consider again the $i$-$T$ diagram in fig. 5. In the lower part of the figure all the isobars may be seen to intersect in a small region near the boundary curve. This is connected with the inversion curve of the Joule-Thomson effect. To the left of these points of intersection an expansion will no longer lead to a decrease of temperature, but on the contrary the temperature will rise. For instance, a point $B$ may easily be reached as the result of an isenthalpic expansion from $A$, whereas a point $B'$ can be reached by such an expansion only from a point $A'$ with a *lower temperature* than $B'$. Now the calculation of which we have spoken shows that the temperature in the first container, for certain values of $p_2$ and $M$, would give points on the diagram corresponding to $B'$, which cannot be reached by an isenthalpic expansion from a higher temperature. However, the low temperatures in a Linde liquefier are produced by means of such expansions. Therefore a point on the diagram corresponding to $B'$, though it may be stable enough in itself, cannot be attained in the course of this process. Our first calculation failed to show this, because we considered only the stationary state and did not concern ourselves with the preliminary process of cooling.

We may determine by a comparatively simple calculation the boundary line on fig. 8 corresponding to the points on the $i$-$T$ diagram that are just accessible. This line is dotted in fig. 8. A pressure of some 40 atm. and $M = 0.2$ are the usual conditions at which a Linde liquefier works. The minimum amount of work to liquefy a kilogram of air may be read off as

$A = 0.71$ kw.h., i.e. about half as much as without the high pressure circuit but still much greater than according to the Carnot process.

Linde's second device for increasing the efficiency of his liquefier is designed to increase the liquefied fraction. This is effected by lowering the initial temperature at which the compressed gas enters the heat exchanger to about $-48°$ C. by introducing an ammonia refrigerator. In the expression (1) for $\epsilon$ the numerator is increased and the denominator decreased by lowering the temperature at 1 and 3, and in this way $\epsilon$ may be almost doubled. The work needed to drive the ammonia refrigerator is very small compared with the economy in $1/\epsilon$. The result is that a combined Linde liquefier with ammonia cooling and high pressure circuit needs theoretically $A = 0.42$ kw.h. to liquefy a kilogram of air. This is considerably nearer the Carnot figure, though of course the inevitable losses play a considerable part.

Table II gives a survey of the *practical* efficiency of the various Linde types of liquefier, the Carnot cycle being in-

Table II. *Efficiency of Linde Liquefiers*

$p_1 = 200$ atm.

| Carnot cycle | Simple Linde process | Linde process with high pressure | Linde process with NH$_3$ | Linde process with high pressure and NH$_3$ ($-50°$) |
|---|---|---|---|---|
| $Q$   97·1 | 9·5 | 7·3 | 17·3 | 14·0 |
| $\epsilon$   1·0 | 0·08 | 0·07 | 0·20 | 0·17 |
| $A$   0·30 | 2·8 | 1·5 | 1·4 | 0·90 |

cluded for comparison. $p_1$ has been chosen as 200 atm., $Q$ is the heat absorbed by each kilogram of air on its way through the apparatus, $\epsilon$ the liquefaction coefficient and $A$ the work in kilowatt hours needed to liquefy 1 kg. of air. The figures are not very accurate and $A$ is naturally much greater than as given by the theoretical computation.

We shall refrain from a calculation of the efficiency of the Claude and Heylandt liquefiers, which may be carried out along similar lines to that of Linde's apparatus but not quite so simply, and confine ourselves to a short description of the method and a summary of the results.

As this method is characterised by almost isentropic changes of state, it is illustrated better on the $T$-$S$ diagram than on the $i$-$T$ diagram. The essential feature of the Claude and Heylandt liquefiers is the expansion machine, in which the air is cooled by an almost adiabatic process. In Claude's first experiments the air was partially liquefied in the cylinder itself at the end of the stroke. This arrangement has, however, several disadvantages. Apart from the fact that the sudden contraction connected with the formation of a liquid phase is apt to lead to explosive effects in the cylinder, the efficiency of an expansion engine at very low temperatures is comparatively small owing to the peculiar shape of the isothermals in the critical region. Moreover, the anomalously high specific heat of a gas near the critical point diminishes the efficiency of the heat exchanger so that under certain conditions it is impossible to emit the expanded unliquefied gas at room temperature. A part of the cold is therefore lost by ejecting cold air into the atmosphere. In view of these difficulties Claude finally refrained from

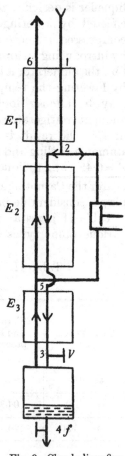

Fig. 9. Claude liquefier.

expanding to complete saturation and covered the last lap by isenthalpic expansion of a separate circuit of air behind the expansion machine. Fig. 9 shows the Claude-Heylandt system in the general form now usually employed. In fig. 10 the process is illustrated on the $T$-$S$ diagram. As in point of fact the expansion in the cylinder is not completely adiabatic, the line 2–5 has been drawn somewhat curved and not vertical.

The efficiency of this type of plant depends in rather a complicated way on a number of factors: the fraction $M$ expanded in the throttle valve, the temperature $T_2$ at which $(1 - M)$ enters the cylinder and the primary pressure $p_1$. The empirical

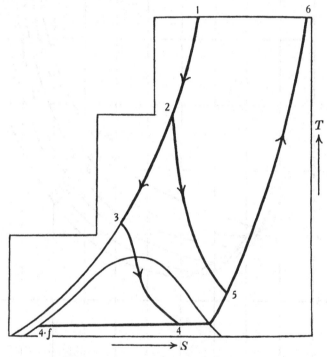

Fig. 10. Claude process on $T$-$S$ diagram.

curves plotted in fig. 11 show the work needed to liquefy a kilogram of air as a function of the fraction $M$ for various values of $p_1$. The air is here supposed to enter the expansion cylinder at the temperatures most favourable for the given values of $p_1$ and $M$, which have also been determined empirically.

The curves demonstrate how the efficiency increases with increasing values of $p_1$ if $M$ is simultaneously increased. For practical reasons Claude works at $p_1 = 40$ atm., choosing

$M = 0.2$ in compliance with the minimum on the corresponding curve. In this case the air should enter the cylinder at $-82°$ C. Heylandt, on the other hand, has chosen $p_1 = 200$ atm. and $M = 0.4$ and has thus gained two advantages. First, as may be

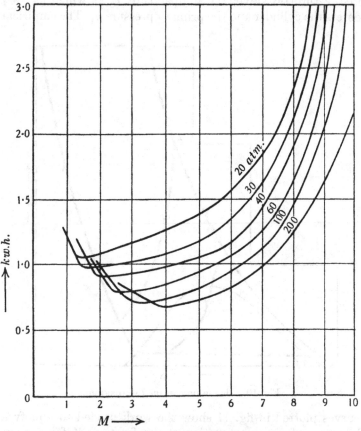

Fig. 11. Efficiency of the Claude-Heylandt process.

inferred from the figure, the efficiency is somewhat greater than in Claude's case, and secondly the most advantageous temperature $T_2$ may be shown to be room temperature. In the Heylandt plants the air therefore enters the expansion cylinder at room temperature, and is there cooled to about $-130°$ C. The heat exchanger $E_1$ is therefore omitted. There appears to be no point in raising the pressure above 200 atm., as this would bring the optimum of $T_2$ to still higher temperatures.

In actual principle the efficiencies of the Linde, Claude and Heylandt liquefiers differ but slightly, the differences being due less to the principles than to technical perfection of design.

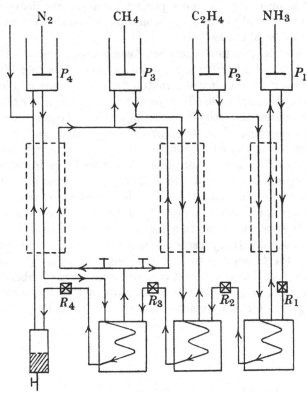

Fig. 12. Cascade according to Keesom.

The best Claude liquefiers need about 0·9 kw.h. for each kg. of liquid air, that is to say, practically exactly the same as the Linde plant with high pressure circuit and ammonia refrigerator. The Heylandt system gives slightly better results, $A$ varying from 0·7 to 0·8 kw.h. per kg.

The essential advantage of the Linde method is that no movable parts of the apparatus are at low temperatures. This greatly simplifies the construction and outweighs many theoretical shortcomings. Claude was confronted by immense difficulties in constructing an expansion machine to work at or near the temperature of liquid air. These difficulties are described very well in his book and were concerned especially with the problem of lubrication. On the whole the Heylandt plant is probably the most perfect piece of machinery and appears to combine most of the advantages and obviate most of the faults of the two principles.

It came as a great surprise when Keesom recently showed that as regards theoretical efficiency the methods of neither Linde, Claude nor Heylandt are a match for the old one of Pictet. Keesom computed the efficiency of a nitrogen liquefier of four cycles: ammonia, ethylene, methane and nitrogen. By judicious selection of the right pressure in each cycle the work required to liquefy a kilogram of nitrogen can be brought down to 0·54 kw.h., allowing for the average heat losses in the heat exchangers and evaporators and for the usual losses in the four compressors. The fact that heat is absorbed in successive stages approximates this plant to an isentropic process. By a very illustrative calculation Keesom demonstrated that the only material entropy losses are incurred in the four expansion valves, the losses due to heat exchange between substances brought together at different temperatures being of subsidiary importance. Fig. 12 shows a diagram of this hypothetical aggregate, which may yet be destined to reassert its priority.

# CHAPTER III

# THE PRODUCTION OF LOW TEMPERATURES

### I. III. 1. *Leiden and Large-Scale Cryogenic Technique*

The inventions of Linde and Claude form a landmark in cryogenic history from which two roads diverge. One we may term Low-Temperature Physics, the other Liquid Gases and their Application. Low-temperature physics is concerned with the properties of matter at low temperatures; in a broader sense it is interested in all properties of matter in so far as experiments carried out at low temperatures may throw some light upon them. Though the production of low temperatures is in itself a domain of low-temperature physics, it is in the main a preliminary step towards carrying out some other research. But nevertheless the very intimate connection which always has and always will exist between low-temperature production and low-temperature research has a very profound significance, to explain which is one of the principal objects of this book. We have now come to a point which may suitably illustrate this connection.

Three years after Linde's first liquefier was successfully operated, the liquefaction of hydrogen was performed by Dewar using the same principle, and ten years later, in 1908, helium was liquefied by Kamerlingh Onnes with the Linde method. We must dwell shortly on the significance of these events.

Hitherto the production of low temperatures has been considered as identical with the liquefaction of gases, and in fact until a few years ago no other method of lowering the temperature had ever been seriously contemplated. This is now easy to explain. We have seen in the last chapter that the liquefaction of gases on a large scale was made possible by utilising the fact that entropy depends on pressure as well as on temperature. In this way it is possible to get rid of entropy with the help of work. In the case of gas liquefaction the gas itself is used as a medium for removing entropy. It is the gas itself which is compressed and expanded, the same gas that is afterwards tapped off in the liquid state. And therefore it is the compression and expansion of this same gas which is responsible for the cooling of the apparatus itself and of any objects which we may choose to immerse in the liquid. But we saw that the amount of cold necessary for the liquefaction of a quantity of gas could not be obtained by simple isothermal compression followed by adiabatic expansion of the same quantity, owing to the high pressures to which the gas would have to be compressed. However, these pressures would have to be far greater if any other substance apart from a gas were used. For to remove a given quantity of heat a definite amount of work is necessary, and the advantage of a gas lies in its compressibility, which enables a large amount of work to be done with a comparatively small change of pressure. Thus when our stock of gases was exhausted at temperatures of about 1° K., the production of still lower temperatures appeared to be faced by insurmountable difficulties. We shall see in Part III how these difficulties were eventually overcome.

The connection between low-temperature production and gas liquefaction is now plain. We have still to demonstrate the connection with low-temperature research. We have named Part I of this book "Phase Equilibrium" and in it we are describing the phenomena attendant on liquefaction and solidification. We are showing how systematic research on the properties of real gases and their equilibrium with liquids has enabled us to liquefy these gases, and how liquefaction has similarly enabled us to determine the physical properties of the liquids and how they depend on the temperature. Now so long as such properties exist that depend on temperature we can make use of these properties to change, that is in our case to lower, the temperature. This is a very general fact and is merely a rather sophisticated way of expressing one of the most popular objects of science, which is to make the laws of nature subservient to our needs.

As Low-Temperature Physics specialises neither in the type of matter on which it carries out its research nor in the particular properties of matter which it investigates, but merely in its methods of approach, the experimental technique which it employs has developed along rather different lines than that used in other domains of physics. Low-temperature technique is faced with three fundamental problems. First, the low temperatures must be *produced*, secondly, they must be *measured*, and thirdly, such apparatus must be devised as will allow us to bring the object to be investigated to the temperature required, to maintain it at that temperature during the experiment and to carry out the experiment under these conditions. The third problem we shall term the *utilisation* of low temperatures.

To judge from the foregoing, the technical apparatus needed for cryogenic work is considerably greater than that usually present in physical and physico-chemical laboratories. In order to carry out experiments at all low temperatures not only large quantities of liquid air are required but, moreover, hydrogen and helium liquefiers equipped with the necessary compressors, pumps and gasometers. These pieces of apparatus require a considerable staff of trained mechanics. Until recently, low-temperature research interested only a relatively small circle of scientists. It is therefore natural that cryogenic work, coupled as it is with considerable technical difficulties, became concentrated in a few large laboratories specially fitted out with the necessary appliances.

By far the most famous of these laboratories is the Natuur-kundig Laboratorium at Leiden in Holland, now known as the Kamerlingh Onnes Laboratory after its great founder. When Kamerlingh Onnes began to equip his laboratory, the principal interest in low-temperature work was presented by van der Waals' Law of Corresponding States. Kamerlingh Onnes' principal object was to study this law at all possible points, low temperatures being a natural feature of the experiments. The systematic nature of the founder and of all the work that bears his name has led to the development of a laboratory rather different from the average place of research. Onnes was not the man to push through sensational experiments in a hurry. There is in fact little sensational information in the *Communications from the Physical Laboratory of Leiden*, a periodical which has become the bible of low-temperature physicists. It is a chronicle of laborious experiments and systematic thought and not always pleasant reading. Yet if we sometimes wonder at the number of years that elapse between an idea and its materialisation, it is well to remember that all the data on matter at low temperatures, now used freely as the common possession of science, were first procured by the experiments of the Leiden school. Moreover, as compared with what Leiden has produced, the gifts of other cryogenic laboratories to science are still almost negligible.

The Leiden laboratory as it stands to-day, equipped with powerful liquefiers for air, hydrogen and helium and with apparatus for work in all domains of physics, may be said to have mastered the three problems of cryogenic technique. Every temperature from $0°$ C. to about $0.7°$ K. may be produced, measured and maintained constant for long periods, and almost every form of experiment may be performed at these temperatures. Not only the Leiden physicists but guests from all parts of the world, themselves not specialists in low-temperature work, can avail themselves of the facilities offered by the Leiden laboratory. These facilities have to a great extent been brought about by the constant attention paid by Onnes and his collaborators to what we have termed the utilisation of low temperatures. In this respect we are confronted on a much larger scale with the same argument as that used by Olszewski: What is the good of producing a low temperature for a few seconds in the interior of a complicated apparatus? Olszewski's demand for a few cubic centimetres of liquid "boiling quietly in a test-tube" is superseded by Onnes' standard of a litre of cold liquid in a Dewar vessel, i.e. practically not boiling at all.

In 1892, three years before Linde's invention, Onnes constructed an air liquefier on Pictet's principle. This is the only case in which this method was made to give comparatively large quantities of liquid air, and it is characteristic of Onnes' work that this plant, though its principle had long been replaced by other methods, was so perfectly constructed that it was retained in constant use till well after Onnes' death in 1924.

Similarly characteristic of Onnes is the fact that, whereas

Dewar first liquefied hydrogen in 1898, it was not until 1906 that liquid hydrogen was produced at Leiden. Onnes was determined not to set up a toy. His liquefier was to be a genuine machine, destined to supply a generation of physicists with as much hydrogen as they needed, and, more important still, to lead to the production of liquid helium. Thus it came about that only two years elapsed between the completion of the hydrogen plant and the liquefaction of helium in 1908.

The liquefaction plants for hydrogen and helium constructed at Leiden, and in the other large cryogenic laboratories which have been installed more recently in various countries, are typical of large-scale laboratory apparatus. Their object was to produce large quantities of liquefied gases and comparatively little attention was paid to efficiency. They are

Table III. *Operating conditions for hydrogen and helium liquefiers of the Linde type*

| Gas | $T_1$ | $p_1$ |
|-----|-------|-------|
| $H_2$ | 70 | 160 |
| He | 15 | 35 |

therefore expensive and can only be afforded in a large laboratory. Until recently they were all based on the Linde principle for the reasons mentioned in the last paragraph. As this method has been sufficiently discussed, we shall here refrain from a detailed description of the apparatus, more especially as it has been exhaustively treated in textbooks. It contains no essentially new ideas, and the preliminary cooling with liquid air and liquid hydrogen is merely a technical complication. The table above gives the usual operating pressures and temperatures as determined empirically and from the inversion curve. The liquefaction coefficients are similar for hydrogen and helium under these conditions and are approximately 0·16.

$T_1$ is the temperature at which the gas enters the heat exchanger, and is obtained by lowering the vapour pressure over liquid nitrogen and hydrogen respectively as far as is consistent with the large volume and the considerable quantities of gas passing through the apparatus. $p_1$ is the most profitable pressure at which the compressed gas should be introduced.

Whereas hydrogen liquefiers have to a certain extent been developed as commercial machines, a helium plant is still a typical piece of laboratory apparatus. This is illustrated by figs. 1 and 2, which show respectively a hydrogen plant made by the Dutch firm Hoek and a helium liquefier built by Linde according to Meissner's construction. Both are in the Low Temperature Laboratory at Kharkov.

The following are the chief points which must be observed in the construction and operation of these liquefiers:

(1) The gases must be exceedingly pure, or else the impurities will solidify and block the expansion valve.

(2) In view of the great temperature drop between the warm and cold ends of the apparatus, and still more owing to the low heats of evaporation of hydrogen and especially helium, very great attention must be paid to heat insulation.

(3) Great care must be taken to avoid explosions when working with compressed and liquid hydrogen. Explosions frequently occur without the presence of a flame through sparks from electric charges formed by friction.

(4) Owing to the danger of explosions from hydrogen and the scarcity of helium it is of importance to prevent the gas from leaking out into the atmosphere. This entails special precautions in the construction of the compressor and of the joints and tubing.

In recent years two very essential novelties have been introduced into the liquefaction of hydrogen and helium by Kapitza at Cambridge.

The necessity of working with exceedingly pure hydrogen had long been felt a serious disadvantage of the Linde method. In order that a hydrogen liquefier may work for several hours on end without the valve becoming blocked, a purity of about 99·8 per cent. is required. The purification of large quantities of gas is a very serious matter, and thus Kapitza's liquefier, which is shown in diagram in fig. 3, and

Fig. 2.
Helium liquefier.

Fig. 1.
Hydrogen liquefier.

in which only a small quantity of pure hydrogen is needed, should be considered as an important step. Kapitza uses a small portion of pure hydrogen, which circulates continuously in his apparatus as a cooling agent, with the help of which large quantities of impure hydrogen in a separate circuit may be successively liquefied. As the second circuit contains no expansion valve, ordinary technical hydrogen may be employed without fear of stoppages. In fig. 3 the pure hydrogen enters at 1, passes through two heat exchangers $A$ and $D$, between which is the liquid nitrogen bath $B$, is expanded at $E$ and partially liquefied in $F$. When $F$ is half full, the hydrogen passes through a spiral cooling condenser $G$ and then out through $D$ and $A$. Ordinary technical hydrogen is admitted at 3 and passes through $A$ and $B$ direct to $G$, where it is liquefied at 3 atm. $G$ is continually drained through tube 4, passing in a spiral through $F$, where it is further cooled to the normal boiling point. From 4 it is tapped to the Dewar vessel, which is thus continuously filled during liquefaction.

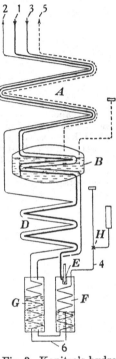

Fig. 3. Kapitza's hydrogen liquefier.

A new stage was reached when Kapitza adapted the Claude method for the liquefaction of helium. The great engineering feat in this innovation was to make an expansion engine work at temperatures below those of liquid hydrogen. This was accomplished by a number of very clever devices. As no grease can be employed at these temperatures Kapitza left a sufficient gap between the piston and the cylinder to avoid friction. By accelerating the compression stroke the amount of gas thus escaping could be made almost negligible. The great advantage of this method is that no liquid hydrogen is necessary. Liquefaction can be attained by precooling with liquid nitrogen alone. In point of fact no precooling at all is needed theoretically, but to work from room temperature would require uneconomically large dimensions for the apparatus. Kapitza's helium liquefier is reproduced in fig. 4.

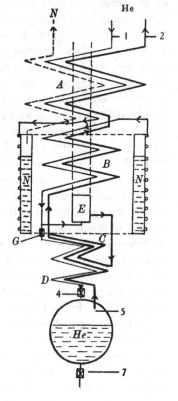

Fig. 4. Kapitza's helium liquefier. 1, helium inlet; 2, helium outlet; A, B, C, D, heat exchangers; N, nitrogen bath and outlet; E, expansion engine; G, reducing valve; 4, throttle valve; 5, low pressure helium; 7, liquid helium tap.

## I. III. 2. *Cryostats*

According to Kamerlingh Onnes the first step in the utilisation of low temperatures must be a device with which they may be maintained constant throughout long periods of time; for only then can accurate measurements be effected. An apparatus in which a low temperature may be maintained constant is termed a cryostat. The Leiden standard for constancy during long periods is 0·01° and the cryostats developed in Leiden are typical pieces of apparatus for a large laboratory in which liquid gases are available in large quantities. Wide intervals of temperature may be bridged and maintained by allowing liquefied gases to boil at various pressures. In Table IV, which illustrates this fact, $T_c$ and $p_c$ are the critical temperature and pressure of the gas named in the first column, $T_b$ is the boiling point at atmospheric pressure, $T_t$ and $p_t$ respectively the temperature and pressure of the triple point and $T_m$ the lowest temperature obtained under normal working conditions with the liquefied or solidified gas.

## Table IV

| Gas | $T_c$ | $p_c$ atm. | $T_b$ | $T_t$ | $p_t$ cm. Hg. | $T_m$ |
|---|---|---|---|---|---|---|
| $NH_3$ | 405·0 | 112·0 | 239·7 | 195·5 | 4·55 | 223·0 |
| $CO_2$ | 304·2 | 73·0 | . | 216·6 | 388·0 | 173·0 |
| $C_2H_4$ | 282·7 | 58·0 | 169·2 | 103·8 | 0·1 | 120·0 |
| $CH_4$ | 190·7 | 45·7 | 112·0 | 90·5 | 8·74 | 87·0 |
| $O_2$ | 154·4 | 49·7 | 90·1 | 54·1 | 0·1 | 56·0 |
| $N_2$ | 126·1 | 33·5 | 77·3 | 63·1 | 9·289 | 50·0 |
| Ne | 44·8 | 29·9 | 27·17 | 24·57 | 32·35 | 24·0 |
| $H_2$* | 33·3 | 12·8 | 20·40 | 13·9 | 5·4 | 9·0 |
| He | 5·2 | 2·5 | 4·19 | . | . | 0·7 |

* See Part III, chap. I.

The figures in the last column have no particular physical significance. In some cases lower temperatures could probably be obtained with the gases in question. We have taken the lowest temperatures that have to our knowledge actually been obtained under normal conditions in the laboratory. The table discloses the ominous fact that between nitrogen and hydrogen and between hydrogen and helium unpleasant spaces are left which cannot be bridged with the help of liquid gases. We shall return to this very shortly.

The simplest form of cryostat for the temperatures covered in Table IV is a cylindrical Dewar vessel fitted with a metal cap by means of an airtight rubber collar. The cap is pierced by numerous metal tubes, one of which is connected to a vacuum pump; one tube admits the object to be studied and usually two are occupied by a mechanism for stirring the liquid. Four conditions must be fulfilled to secure a constant temperature. The inflow of heat from without must be reduced to a minimum, the vapour pressure must be maintained rigidly constant, heat gradients within the liquid must be prevented and the liquid itself must be exceedingly pure. Impurities lead to a gradual upward trend of the temperature, since the component with the highest vapour pressure evaporates first. A well-silvered glass Dewar vessel usually suffices to reduce the heat inflow, but when the vessel contains liquid hydrogen it is advisable to surround it with a second vessel containing liquid air. In working with liquid helium three concentric Dewar vessels are employed; the innermost contains helium, the second vessel liquid hydrogen and the outer nitrogen, which is used in place of liquid air to reduce the danger of explosions. In order to keep the pressure constant, a differential manometer filled with oil is used apart from a mercury manometer, and the pressure is regulated by hand with a sensitive valve. In some cases automatic pressure regulation has been successfully attempted. The stirring mechanism is exceedingly important, especially in liquid oxygen, the heat conductivity of which is low, whereas its viscosity is great. Unless great precautions are observed, temperature differences of several degrees exist between the surface of the liquid and the bottom of the Dewar vessel. In the Leiden construction two half-cylinders of German silver are fitted with light vanes capable of pivoting through 90°. The half-cylinders are alternately raised and lowered by means of an electromagnet, the pressure of the liquid raising and thus opening the vanes when the stirrer is lowered and closing them in the next half period. A stirrer can, of course, only be employed above the triple point, so that for this type of cryostat the temperature intervals covered by Table IV must be considerably reduced. Fortunately the heat conductivity of solid hydrogen is high, so that here tolerable constancy can be maintained even without the use of a stirrer.

The production and maintenance of temperatures not covered by liquefied gases is a far more difficult task. Here some form of gas cryostat is usually employed. The gas is first passed through a Dewar vessel containing a liquid colder than the temperature required, and then slightly heated and allowed to flow through a second Dewar vessel containing the substance to be investigated. As a result of the low heat content of the gas and the absence of two-phase equilibrium, rather elaborate mechanisms have had to be devised, if the temperature is to be maintained accurately constant for long periods. Fig. 5 shows a gas cryostat for temperatures above liquid air. The vessel containing the object to be studied is filled with petroleum ether, which remains liquid down to liquid air temperatures. This method is usually preferred to a cryostat containing liquid ethylene, as ethylene is difficult to purify, rather expensive, and soluble in the oil of the vacuum pumps.

For temperatures between liquid air and liquid hydrogen and between solid hydrogen and liquid helium, various forms of cryostats have been devised, the latter interval being the most difficult to cover. In fact until now only comparatively few accurate measurements have been performed at these temperatures. In the course of the last few years the Leiden physicists have constructed a number of cryostats for this interval, for the details of which we must refer the reader to the Leiden Communications.

The cylindrical glass Dewar vessel, as used in the Leiden cryostat, silvered to minimise heat inflow from radiation with two narrow strips unsilvered to allow observation, has left its mark on almost all types of cryogenic work. A very large percentage of low-temperature experiments has been carried out in such vessels, and one of the chief difficulties of physicists is to fit pieces of apparatus together so that they may be enclosed in such cryostats. In some types of experiment, such as calorimetry, electrical conductivity measurements, etc., this is easy enough, but in other cases, especially when large auxiliary apparatus is required, which cannot be separated from the object to be investigated by glass walls, it is almost impossible, and the history of low-temperature technique is a permanent war between the glass Dewar vessel and other pieces of apparatus. This war has not yet been decided and, in fact, the problem whether the glass Dewar will in time be superseded by new devices, numerous exceedingly clever suggestions for which have at various times been put forward, bears a very different appearance in the Leiden laboratory on the one hand and in the small cryogenic laboratories on the other.

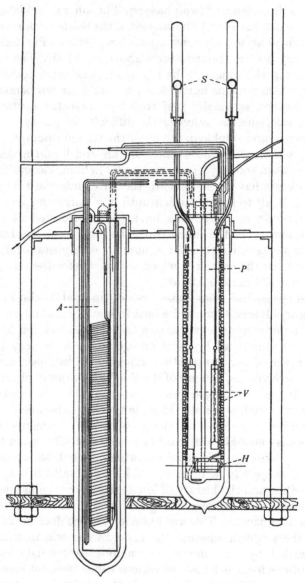

Fig. 5. Gas cryostat. *A*, Dewar vessel for precooling; *P*, petroleum ether; *V*, stirring vanes; *H*, heater; *S*, iron spheres for magnetic stirrer.

## I. III. 3. *Small-Scale Cryogenic Technique*

As the results of cryogenic work began to attract attention, numerous scientists, working in all fields of physics, became desirous of carrying out certain experiments at low temperatures and, not being in possession of the complicated and expensive apparatus in use in the great laboratories, decided to do without them. It is clear that the problems sketched above were faced in quite a different manner in these two types of laboratories, and, in fact, a number of problems were investigated and solved in the small laboratories which would never have occurred to scientists in the old cryogenic institutes, where liquid gases are present in sufficient quantities.

In very many experiments the object to be investigated at low temperatures is small and thus the heat to be removed from the object in order to cool it is negligible as compared with that required to liquefy a litre of gas, and it appeared evident that simpler methods could be devised. Various types of laboratory liquefiers for small quantities of gases were a step in this direction. The line of thought which led to the second step is shortly as follows. Low temperatures may be roughly classed in three groups: temperatures obtained with the help of liquid air, those to be reached with liquid hydrogen, and those for which helium must be liquefied. Liquid air is available almost everywhere, liquid hydrogen in quite a number of centres, liquid helium in very few places indeed. It was therefore desirable to place the temperatures of liquid hydrogen within the reach of scientists possessing liquid air, and the temperatures of liquid helium within the reach of those working with liquid hydrogen.

An attempt in this direction was first made by Simon in his "Desorption Method", which has lately given rise to considerable discussion. Simon considered his method as a revival and continuation of Pictet's cascade. As shown in Table IV the series of gases available for the cascade ceases at nitrogen. With the help of liquid nitrogen boiling at low pressure it is impossible to reach the critical point of hydrogen. The only gas whose critical temperature lies between those of nitrogen and hydrogen is neon. But even this temperature—42° K.—

is well beyond the reach of nitrogen and oxygen; moreover, neon is far too scarce as yet to be suitable for this purpose. Similarly, helium temperatures cannot be attained with the help of liquid or solid hydrogen, and no intermediate substance exists. Between hydrogen and helium temperatures Simon thereupon introduced an "artificial" intermediate substance in the form of helium sorbed on charcoal. It is a well-known fact that gases are readily adsorbed on charcoal at temperatures not very far removed from their critical points, and it is also known that the heat of adsorption is very considerable. Simon's apparatus, which is shown in fig. 6, consists of a small copper vessel $A$ filled with charcoal, surrounded by a vacuum jacket $R$ suspended in a Dewar vessel $D$ filled with liquid hydrogen. At the beginning of the experiment $R$ is filled with helium at low pressure to enable the temperature of the hydrogen in $D$ to be communicated to the interior. The vapour pressure in $D$ is then lowered as far as possible, and simultaneously helium gas is admitted to the charcoal in $A$ and sorbed in large quantities, the heat of sorption being communicated to the liquid hydrogen in $D$ through the gas in $R$. Thereupon $R$ is evacuated and $A$ thus insulated from $D$. Finally the sorbed helium is pumped off with a powerful pump. As a result of the adiabatic insulation, the heat of adsorp-

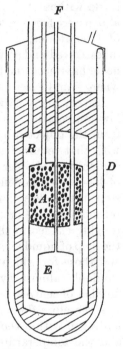

Fig. 6. Simon's desorp-tion apparatus.

tion has now to be carried by the charcoal and the vessel $A$. The temperature in $A$ thus rapidly falls and the boiling point of helium may be reached. It is then possible to liquefy some helium through the tube $F$ into the vessel $E$, if required. However, this is usually not the aim of the experiment, and as a rule $E$ will serve to contain the object to be investigated. In the apparatus here sketched the initial adsorption pressure before evacuation did not exceed 1 atm. In a later type developed by Mendelssohn in Simon's laboratory, adsorption pressures of 5 atm. were employed. Apart from the greater heat of sorption thus available, this has the advantage of making use of the heat of expansion of the gas in $A$, which plays a certain part in the process. In Mendelssohn's experiments the temperature of the hydrogen in $D$ was 10° K. Desorption and expansion from 5 to 1 atm. produced a temperature of 7° K. By lowering the sorption pressure to about 0·1 mm. Hg., temperatures rather lower than 4° K. were obtained. Helium was thereupon condensed in the apparatus, and by reducing the pressure the temperature was lowered to 1·5° K. An apparatus of this kind evidently possesses numerous advantages. To begin with it is comparatively cheap and simple. Very small quantities of helium gas are needed and no high pressures are required. Secondly, only a small portion of the heat of sorption is actually needed to cool the object to the temperature of liquid helium; the rest can be used to maintain the temperature against heat inflow from without. Temperatures below 4° K. can be kept thus for several hours. For after the lowest temperature has been reached, considerable quantities of helium remain on the charcoal. But probably the chief advantage of the method is that the temperatures under the command of the apparatus are not bound to the boiling point of helium. By varying the amount of gas adsorbed or, which comes to the same thing, the adsorption pressure, the lowest point reached, and thus the temperature ultimately maintained in the apparatus, can be adjusted to any point between the boiling points of hydrogen and helium, and it is thus possible to work in this interval, which is inaccessible where only liquefied gases are employed. Simon's artificial link between hydrogen and helium in fact corresponds to an infinite series of substances due to the free variable given by the amount of gas sorbed.

The intermediate temperatures between liquid air and liquid hydrogen and between liquid hydrogen and liquid helium may be reached comparatively simply and maintained reasonably constant with a miniature Linde apparatus introduced by M. Ruhemann and shown in fig. 7. It consists of a small heat exchanger, a narrow copper tube $R_1$ inside a slightly wider tube $R_2$ of German silver, wound to a spiral and enclosed in a cylindrical vacuum jacket $A$. The expansion valve $D$ is a small brass screw adjusted from outside the "liquefaction vessel" $B$ so as almost to block the exit of the interior copper tube. When the screw has been adjusted, it is lightly soldered to the wall of $B$ with Wood's metal in order that a vacuum may be maintained in $A$, the external cylinder of which is then soldered. To $B$ may be attached the object to be studied or any parts that are to be cooled. The entire apparatus, which need be no more than some 30 cm. long and 4 cm. in diameter, is enclosed in a glass Dewar vessel $G$, which is filled with liquid air or hydrogen, according to whether the interval between air and hydrogen or that between hydrogen and helium is to be attained. In the first case hydrogen gas is circulated through the heat exchanger, in the second case helium. As not more than 2 litres of gas need flow per minute, it may be taken straight from the flask, and in the case of helium may be compressed back at leisure. In the case of hydrogen it is generally not worth while to recover the gas, which may simply be conducted out of the window. By varying the rate of flow and the initial pressure, any temperature can be maintained down to the boiling point, the apparatus being so adjusted that the Joule-Thomson effect just compensates the heat leakage. It is even easier to adjust the pressure so that small quantities of gas are liquefied in $B$. With dimensions as given above and 120 atm. of hydrogen flowing at an average rate of 2 litres a minute, liquefaction begins after about 8 minutes. Helium at 40 atm. may be liquefied under the same conditions with liquid hydrogen in $G$ in the same period of time. The principal disadvantage of this apparatus as against the desorption method is that very pure gas must be employed, all impurities solidifying and thus blocking the valve. The chief advantage is that it is a truly

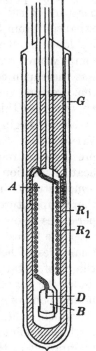

Fig. 7. Miniature Linde apparatus.

continuous process and may in principle be prolonged indefinitely, whereas the desorption process naturally ceases when no more gas is left on the charcoal.

By far the simplest form of laboratory liquefier, which appears destined to play a considerable part in popularising low-temperature physics, is also due to Simon. The method is in the main a repetition of Cailletet's experiment under far more congenial conditions. Helium is compressed to about 100 atm. in the small copper vessel $B$ (fig. 8), the heat of compression being conducted to the liquid hydrogen in the Dewar vessel. Thereupon the jacket $Z$ is evacuated and the helium pressure relaxed through a German silver capillary. The helium is cooled just as in Cailletet's first experiment, and after the expansion half the vessel $B$ is left filled with liquid helium. The reason why this simple method is particularly suited to very low temperatures is that the specific heat of all solid bodies, in this case the walls of $B$, is so small at these temperatures that in reality only the helium itself absorbs any appreciable heat. The cold gas emitted through $R$ is of course lost in this process, but in

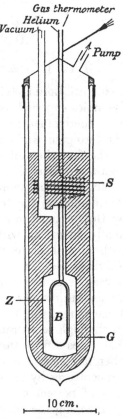

Fig. 8. Simon's expansion apparatus.

these small dimensions heat economy plays no part at all. The liquid helium in $B$ can be kept for an hour or more, which is sufficient for quite a number of experiments to be performed.

# CHAPTER IV

# THE MEASUREMENT OF
# LOW TEMPERATURES

### I. IV. 1. *Gas Thermometry and the Kelvin Scale*

Not until Lord Kelvin showed that the second law of thermodynamics supplied us with an absolute scale of temperature
was a thermometer anything more than an arbitrary definition
of a zero point and a degree.

The fact that all gases at sufficiently
high temperatures and low pressures approach more and more closely to "ideal"
thermometers, the temperatures defined
by their expansion coefficients becoming
finally identical with each other and
with the absolute temperature of the
Kelvin scale, has established the gas
thermometer as the primary instrument
for measuring all temperatures at which
it can be realised. This interval is limited
on the high temperature side by the fact
that gases diffuse through all solid walls
when hot enough, and on the side of low
temperatures by the deviations from the
laws of perfect gases. At temperatures
approaching the critical, Boyle's law can
no longer be used to define the temperature of a gas from its volume at normal
pressures.

The gas-thermometer principle may
best be illustrated by a very simple model,
occasionally employed by Simon for rough

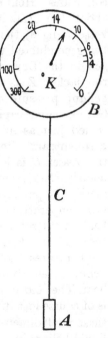

Fig. 1. Simple gas
thermometer.

measurements. A bulb *A*, which is at the low temperature
to be measured, is connected by means of a narrow capillary
tube to an ordinary vacuum gauge *B* (fig. 1). The system is
filled with helium at room temperature to give a pressure of

1 atm. At a given temperature $T$ at $A$ the vacuum gauge, which remains at room temperature $T_0$, will show a pressure $p$ depending on the relative volumes of $A$ and $B$. If these volumes are $v$ and $v_0$ respectively, Boyle's law will give

$$p = \frac{nRT}{v} = \frac{n_0 RT_0}{v_0},$$

where $n_0$ and $n$ are the number of gramme-molecules in $B$ and $A$ respectively. We assume that the volume of the capillary may be neglected. Now

$$n + n_0 = N = \frac{p_0 (v + v_0)}{RT_0},$$

where $p_0 = 1$ atm. Thus

$$\frac{pv}{RT} + \frac{pv_0}{RT_0} = \frac{p_0 (v + v_0)}{RT_0}.$$

Whence

$$p \left( \frac{v}{RT} + \frac{v_0}{RT_0} \right) = \frac{p_0 (v + v_0)}{RT_0}.$$

Therefore

$$p = \frac{p_0 (v + v_0) T}{vT_0 + v_0 T}.$$

To illustrate this relation let us put $p_0 = 1$ atm., $v = v_0 = 1$ c.c., $T_0 = 300°$ K. Then

$$p = \frac{2T}{300 + T}.$$

At low temperatures this is a sensitive thermometer giving almost linear readings, but in the neighbourhood of room temperature it is far less sensitive.

This primitive model brings out in caricature all the principal properties of a gas thermometer. Every normal gas thermometer consists of a vessel of gas at the temperature to be measured, connected by a capillary to a measuring vessel at room temperature. In order that the gas pressure may be proportional to temperature, the volume $v_0$ of the measuring vessel must be small as compared with that of the bulb. Moreover, when the dimensions of the two vessels are given, the readings will be the more nearly linear, the lower the temperature. For at low temperatures nearly all the gas is concentrated in the bulb.

As according to Boyle's law the pressure and volume coefficient of a gas are identical, it is of no significance if we measure the gas pressure at constant volume instead of the volume at constant pressure. In practice the constant volume gas thermometer, as illustrated in fig. 2, has proved superior. Its chief features are the gas bulb $B$ and the mercury manometer $M$, connected to the bulb by the capillary $C$. The manometer is so constructed that the meniscus in the shorter tube $T_1$ is maintained at a constant level near the entrance of the capillary at $K$. In order that this may be effected at varying pressures of the gas in $B$, mercury must be admitted to or removed from the manometer through the stop-cocks $t_1$ and $t_2$ respectively, the former being connected to a mercury reservoir $R$. The final adjustment is brought about by the screw $s$, which presses on a steel membrane $m$, thus slightly increasing or decreasing the volume of the mercury vessel. The pressure is then read from the meniscus of the mercury in $T_2$, either with the aid of a cathetometer or with a simple magnifying glass and nonius.

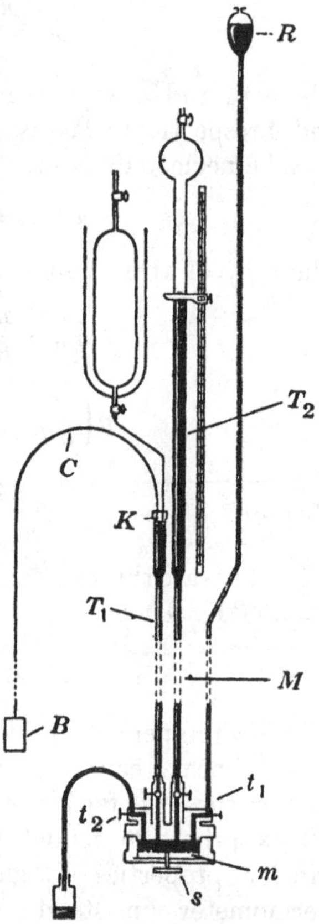

Under ideal conditions the temperature would be determined by the equation

Fig. 2. Constant volume gas thermometer.

$$t_p = \frac{(p_t - p_0) \cdot 100}{p_{100} - p_0},$$

where $p_0$, $p_{100}$ and $p_t$ are the pressures at 0°, 100° and $t$° respectively. In reality a correction must be applied for the "noxious volume", consisting of the capillary $C$ and the volume of gas above the meniscus in $T_1$. These volumes must therefore be made as small as possible in comparison with $B$. The capillary may usually be neglected, but since the mercury tubes must be fairly wide to give accurate readings, the space above the meniscus can hardly be less than 0·6 c.c.

We have already mentioned that, under normal conditions, not even the so-called permanent gases show identical expansion coefficients, i.e. Boyle's law must always be considered as but an approximation to truth. Thus, apart from the practical corrections to be applied to gas-thermometer readings, another set of corrections must be considered, which are due to the deviations from the perfect state. However, since the gas thermometer is to be a primary instrument for measuring temperatures, i.e. is to *define* temperatures, these corrections offer considerable fundamental difficulties. For we have no means of calibrating a gas thermometer except with other gas thermometers.

The reduction of the readings of a gas thermometer to the absolute scale is equivalent to determining the equation of state of the gas, which connects pressure and volume with temperature. Neither Boyle's law nor van der Waals' equation is sufficiently accurate to enable us to compute temperature from volume or pressure readings, and the same applies to all the other so-called theoretical equations of state that have been put forward from time to time. It was for this reason that Kamerlingh Onnes expressed the equation of state as an infinite series containing as many arbitrary constants as appeared necessary. This empirical equation reads as follows:

$$pv = A\left(1 + B/v + C/v^2 + \ldots\right), \qquad \ldots\ldots(1)$$

where $A$, $B$, $C$, ... are considered as functions of the temperature alone. In most cases the series may be broken off after the linear term. The more accurate the work the more terms must be included. $A$, $B$ and $C$ are determined by measuring isothermals, i.e. the relation between pressure and volume at constant temperature. From a group of isothermals, $B$ and $C$ may be expressed as series in $T$, the true temperature being attained as a result of successive approximations, whereby it is assumed that Boyle's law is accurately valid for the limiting case of $p = 0$.

Onnes himself reasoned as follows: Temperature is defined by the Second Law of Thermodynamics according to the formula

$$dS = \frac{du + p\,dv}{T},$$

or any similar equation. This enables us in principle to determine $T$ by measuring any properties of a body that are connected by the Second Law in equations involving temperature. Now we assume that some gases under certain conditions strictly follow Boyle's law in the form $pv = A$, where $A$ depends on temperature alone. These conditions Onnes defines as the Avogadro State.

Now if we measure temperatures with a thermometer filled with a certain gas under certain conditions, employing (1) as our equation of state, and if we determine $B$ and $C$ by measuring isothermals at various temperatures, we make use of a temperature scale which Onnes terms the Avogadro Scale for a Certain Gas. This scale need not necessarily agree with the theoretical Avogadro Scale, since equation (1) may not be correct, and moreover $B$ and $C$ may not have been determined with sufficient accuracy. The only indication that we have that the two scales agree is that Avogadro Scales obtained with various gases agree with one another.

For practical purposes it has appeared advisable to introduce standard conditions for measuring temperature. For this purpose the Société Internationale du Froid introduced the International Helium Scale, defined by a gas thermometer at constant volume filled with helium at a pressure of 1 m. of mercury at 0° C. This scale can be corrected to the Avogadro Scale for Helium by making use of the values of $A$, $B$ and $C$ as determined at Leiden.

It is only at temperatures below the boiling point of helium that the fundamental difficulties of thermometry begin. For here we approach the limits of the gaseous state. Fortunately the isothermals measured at Leiden have shown that helium, even in the immediate vicinity of the liquid vapour curve, behaves very much as a perfect gas. Thus the quadratic term in equation (1) need nowhere be taken into account and even the term $B/v$ forms a very insignificant correction. However, as the vapour-pressure curve falls very sharply below the boiling point, whereas the pressure in the gas thermometer must be kept below this curve, it is necessary to work at very low pressures, at which the readings of an ordinary mercury manometer are not sufficiently accurate. For this reason Kamerlingh Onnes introduced the hot-wire manometer, as suggested by Knudsen, which makes use of the fact that at low pressures the heat conductivity of a gas is highly dependent on pressure. A fine Wollaston filament is heated by a constant weak current, and the temperature of the filament determined by its electrical resistance. This temperature is *ceteris paribus* a function of the heat conductivity and thus of the pressure of the surrounding gas, and the manometer can be calibrated for a definite gas with a high degree of accuracy.

However, the determination of the pressure is only one of the minor difficulties of low-temperature thermometry. By far the most serious obstacle is the so-called thermomolecular pressure, discovered by Knudsen in 1912. We have seen that in order to minimise the correction due to noxious volume it is necessary to employ narrow capillaries connecting the thermometer vessel at low temperature with the manometer working at room temperature. Now at very low temperatures and pressures, when the mean free path of the gas molecules is no longer small as compared with the diameter of the tube, a considerable pressure gradient exists between vessels at different temperatures connected by a capillary. As long as we have no method of measuring low temperatures independent of the gas thermometer, this pressure gradient or, as it is frequently termed, the Knudsen effect, cannot be determined experimentally. On the other hand, a theoretical computation is only possible for the limiting case, when the mean free path is great as compared with the diameter of the tube. This condition is as a rule not fulfilled, nor is it possible to gauge with sufficient certainty the degree of fulfilment necessary, in order that the formulae may be employed. Kamerlingh Onnes and his co-workers have undertaken numerous experimental and theoretical researches to determine the Knudsen correction, and the temperature scale to about 0·8° K. is now probably correct to within a hundredth of a degree.

## I. IV. 2. *The Vapour Pressure as a Measure of Temperature*

In practice, gas thermometers are seldom employed, as several secondary instruments are far easier to handle. These can be calibrated with gas thermometers or with the help of certain fixed points that can be joined up with interpolation curves. The fixed points employed in low-temperature thermometry are in the main boiling points and triple points of condensed gases and a number of transition points of solids that have been determined with great accuracy with gas thermometers. In principle, boiling points are of course less advantageous than triple and transition points, as they are highly dependent on pressure. However, pressure determination can be carried out very accurately, and calibrations can be made at a whole series of pressures. In choosing thermometrical fixed points, only such substances should be taken as can be obtained in a very pure condition without great trouble, or else only such points that are not very greatly dependent on absolute purity. Transition points are particularly dangerous in this respect, as small impurities may lower them by several tenths of a degree. In Table V is given a list of possible fixed points below zero Centigrade, those in black type being in general use. It is highly probable that several others would be just as useful, but in most cases they have not been measured with sufficient accuracy.

Three types of secondary thermometers have been and still are used in low-temperature work: vapour-pressure thermometers, thermocouples and resistance thermometers. Each have their particular advantages and defects, and in many cases it is a question of dispute which thermometer is the most suited for a definite purpose.

The vapour pressure of a pure liquid as an accurate indication of temperature was first suggested by Stock. The principal advantage of this method is its high sensitivity in certain temperature intervals. The vapour-pressure curves of the so-called permanent gases are very steep in the vicinity of the normal boiling point, so that temperature differences of 1° generally correspond to a pressure difference of a good many centimetres. This fact is at the same time responsible for the

## Table V. *Fixed Points*

| | | Temperature | Observer(s) | Reference |
|---|---|---|---|---|
| $CCl_4$ | Trip. | − 22·87 ± 0·05 | Johnston and Long | J. Amer. Chem. Soc. **56**, 31, 1934 |
| **$Hg$** | Melt. | − 38·87 | International Temperature Scale | |
| $CCl_4$ | Trans. | − 47·66 ± 0·05 | Johnston and Long | Loc. cit. |
| **$CO_2$** | Subl. | − 78·50 | International Temperature Scale | |
| $C_6H_5OH$ | Melt. | − 95·0 | Timmermans | Leiden Comm. Suppl. 64 a |
| $C_2H_4$ | Boil. | − 103·72 | Henning and Stock | Zs. f. Phys. **4**, 226, 1921 |
| $iso\,C_5H_{12}$ | Melt. | − 160·0 | Timmermans | Loc. cit. |
| $C_2H_4$ | Trip. | − 169·4 | | J. Amer. Chem. Soc. **43**, 1098, 1921 |
| **$CH_4$** | Boil. | { − 161·37 | Henning and Stock | Loc. cit. |
| | | 111·57° K. | Thermoelectric Temperature Scale | |
| **$O_2$** | Boil. | − 182·97 | Keesom, v. d. Horst, and Jansen | Proc. Acad. Amst. **32**, 1167, 1929 |
| **$N_2$** | Boil. | 77·33° K. | Thermoelectric Temperature Scale | |
| $CO$ | Trip. | − 204·98 | Clusius | Zs. f. phys. Chem. (B), **3**, 41, 1929 |
| $N_2$ | Trip. | { − 210·02 | Henning | Zs. f. Phys. **40**, 775, 1927 |
| | | − 209·99 | Keesom and Bijl | Physica, **4**, 305, 1937 |
| $CO$ | Trans. | − 211·69 | Clusius | Loc. cit. |
| | | { − 211·65 ± 0·05 | Clayton and Giauque | J. Amer. Chem. Soc. **54**, 2610, 1932 |
| $O_2$ | Trip. | − 219·11 | Giauque and Johnston | Loc. cit. |
| $O_2$ | Trans. | { − 218·81 | Giauque and Johnston | J. Amer. Chem. Soc. **51**, 2300, 1929 |
| | | − 229·44 | | Loc. cit. |
| $N_2$ | Trans. | − 237·8 | Clusius | Loc. cit. |
| $Ne$ | Boil. | 27·16° K. | Crommelin and Gibson | Leiden Comm. 185b |
| $Ne$ | Trip. | 24·57° K. | Crommelin and Gibson | Loc. cit. |
| **$H_2$** [n] | Boil. | − 253·81 | Bonhoeffer and Harteck | Zs. f. phys. Chem. (B), **4**, 113, 1929 |
| [p] | Boil. | − 252·94 | | |
| [n] | Trip. | − 259·25 | Bonhoeffer and Harteck | Loc. cit. |
| [p] | Trip. | − 259·38 | | |
| **$He$** | Boil. | 4·220° K. | Keesom, Weber and Schmidt | Proc. Acad. Amst. **32**, 1314, 1929 |
| $He$ | λ-point | 2·190° K. | Keesom and Keesom | Leiden Comm. 221 e |

Trip. = triple point.
Melt. = melting point at 760 mm. Hg.
Boil. = boiling point at 760 mm. Hg.

Subl. = point of sublimation.
Trans. = transition point.

λ-point, see Part II, chap. II.
n = normal hydrogen.
p = parahydrogen, see Part III, chap. I.

most outstanding defect of these instruments. Each is applicable to only a very narrow temperature interval. Indeed, the lower the boiling point of the substance, the more sensitive the vapour-pressure thermometer and the smaller the number of degrees below the boiling point at which the thermometer ceases to be useful. For when the vapour pressure of the liquid is so low that a McLeod gauge or a hot-wire manometer are needed to determine it, the thermometer looses its simplicity and thus its charm. Moreover, most liquefied gases solidify at appreciable vapour pressures and are then unsuitable as thermometers on account of the low heat conductivity of non-metallic solids, which delays the establishment of equilibrium. Next to their high sensitivity it is their simplicity which has made these instruments popular, especially in physico-chemical laboratories, where highly developed physical technique is frequently lacking. For as the vapour pressure is independent of the amount of substance in each phase, the volume of the vessel containing the liquid, the connecting tubes and the manometer need not be known, so that no calibration is necessary. Since this type of thermometer always shows the temperature at the coldest point of the apparatus, it is necessary that all points in contact with the thermometer tube shall be warmer than the point to be measured.

A rise or fall of the vapour pressure is always accompanied by evaporation or condensation of a certain amount of liquid, dependent on the volume of the manometer and tubing, and thus with the absorption or emission of latent heat. This, of course, gravely impairs the utility of these thermometers for calorimetric work, since the heat given off or absorbed is considerable. Thus the evaporation of only 1 mm.$^3$ of liquid oxygen is associated with the absorption of about 0·1 cal.

It is at very low temperatures that vapour-pressure thermometers have proved most useful, especially at the times when these temperatures were first attained and the gas thermometer corrections were not yet determined with sufficient accuracy. For it is usually not difficult to find a simple interpolation formula connecting vapour pressure with temperature. An accurate theoretical formula can be deduced only when the thermodynamical data of the substance are known, and this is seldom the case. However, when the specific heats of both phases are determined as functions of the temperature as well as the heat of evaporation, the vapour pressure may be employed to compute the absolute temperature directly from the Second Law without having recourse to the gas-thermometer scale. This was actually carried out by Simon and Lange in the case of hydrogen.

In Leiden, the vapour pressure of helium was used for a number of years practically as a definition of temperatures below the boiling point of helium. For it was the only thermometer that was comparatively easy to realise in this region. From a preliminary calibration with a helium gas thermometer an interpolation formula was deduced, which was subsequently successively corrected after a prolonged research had made the gas-thermometer scale more and more reliable.

A curious difficulty is here encountered by the existence at 2·19° K. of a discontinuity in the properties of liquid helium, to which we shall refer again later. The density, compressibility and specific heat of the liquid undergo rapid changes in the neighbourhood of this point, the origin of which is still uncertain. For the present it appears profitable to assume that liquid helium occurs in two "modifications", one, He I, being stable above 2·19° K. and the other, He II, below. The transition temperature is a function of pressure and the equilibrium curve between He I and He II has been traced up to the melting curve (see chap. VI). It is not surprising that a single formula does not hold for the entire vapour-pressure curve. Keesom therefore employs two vapour-pressure formulae, one for He I and one for He II. The following are the latest formulae, which mark the position in 1932:

Above 2·19° K.:
$$\log p = -3 \cdot 024/T + 2 \cdot 208 \log T + 1 \cdot 217.$$

Below 2·19° K.:
$$\log p = -3 \cdot 018/T + 2 \cdot 484 \log T - 0 \cdot 00297 T^4 + 1 \cdot 197.$$

## I. IV. 3. *Electrical Thermometry*

Thermocouples, though very useful for rough determinations, are seldom employed for accurate temperature measurements. In general the thermoelectric forces generated between two joints of different metals fall off rapidly at low temperatures, so that below the boiling point of hydrogen the sensitivity of all thermocouples is small. Moreover, no suitable interpolation formulae have been found to join up calibrations made at fixed points. The three couples used at low temperatures are copper-constantan, iron-constantan and gold-silver. The thermoelectric force of the gold-silver couple has a comparatively simple temperature curve but low sensitivity. Iron-constantan, though sensitive, is not very reliable, since the iron is apt to oxidise, thereby changing the thermoelectric force. Copper-constantan is most frequently employed.

The fact that the thermoelectric force of a thermocouple is a function of the temperatures of both joints is at once an advantage and a disadvantage. It facilitates the measurement of temperature differences, and, especially in so-called adiabatic work (see Part II, chap. II), when two parts of an apparatus are to be maintained at the same temperature, it enables the thermocouple to be used as a zero instrument. On the other hand, it is necessary to keep the warm junction at a well-defined constant temperature, for which in most cases a cryostat of melting ice is employed. In some cases it is preferable to place the warm junction in a cryostat of boiling oxygen or nitrogen.

By far the most important secondary thermometer from about $+500°$ C. downwards is that based on the relation between temperature and the electrical resistance of metals. As we shall see in Part IV the resistance of all pure metals falls off rapidly as the temperature is reduced, and several metals have been used as resistance thermometers at various temperatures. Platinum, which is obtained in sufficient purity fairly easily, is most frequently employed, one of its chief merits being that above $0°$ C. its resistance may be expressed with great accuracy as a quadratic function of the temperature:

$$R_t = R_0 + at + bt^2, \qquad \qquad \dots \dots (1)$$

where $R_t$ is the resistance at temperature $t$, $R_0$ the resistance at $0°$ C. and $a$ and $b$ are constants depending on the degree of purity and on the thermal and mechanical treatment of the sample. At temperatures below $0°$ C. the relation is not so simple and a term of the fourth degree must be introduced, which is, however, almost independent of the specimen. At temperatures below liquid oxygen even this relation no longer holds, and it is necessary to calibrate the resistance thermometer at a number of points obtainable with liquid oxygen and hydrogen, the interval between $55°$ and $20°$ K. being the most difficult to bridge. Below $10°$ K. the resistance of platinum becomes almost independent of temperature, and the lower the purer the sample.

Lead resistances are frequently used, especially at hydrogen temperatures, as they are slightly more sensitive. But the relation between the resistance and temperature is more complicated, and they can be applied only as far down as $7°$ K., since lead here becomes supraconductive (see Part IV, chap. II). At temperatures obtainable with liquid helium, constantan was formerly employed as a resistance thermometer. Though at higher temperatures its resistance is to all intents and purposes independent of temperature, it possesses at helium temperatures a small temperature coefficient, which has the advantage of being practically constant. The constantan thermometer has now given place to phosphor bronze, which was shown by Keesom to have a high temperature coefficient of resistance at very low temperatures, probably caused by slight supraconductive impurities.

Resistance thermometers are in most cases thin wires wound loosely on unglazed porcelain frames. In order that the readings can be well reproducible the wires must be free from strain. If they are tightly wound they will be strained by their own thermal contraction at low temperatures. Moreover, the wires must be annealed at about $500°$ C. and immersed several times in liquid hydrogen before the final calibration takes place.

The most accurate apparatus for measuring the resistance is a potentiometer, and here the resistance of the leads cancels out. It is therefore not necessary to employ a high zero-point resistance for the wire. Platinum wires of a few tenths of a millimetre in diameter are most frequently used, the zero resistance being between 20 and 100 ohms. If a Wheatstone bridge is used, the leads must be measured separately and higher zero-point resistances are needed. This necessitates the employment of very fine wires. On the other hand, the Wheatstone bridge gives the resistance directly as a result of one reading, whereas two readings are required with a potentiometer. This sometimes makes a bridge preferable when rapidly changing temperatures are to be determined.

As the helium thermometer is difficult to realise and still more difficult to transport, the Platinum Scale has been introduced as the international juridical temperature scale in most countries, down to the boiling point of oxygen. Temperatures above $0°$ C. are defined according to formula (1); at lower temperatures by the following formula:

$$R_t = R_0 \{1 + At + Bt^2 + C(t - 100)t^3\}.$$

The internationally recognised fixed points for calibration are:

| | | |
|---|---|---|
| The boiling point of sulphur | ... | $444·60°$ C. |
| The boiling point of water | ... | $100·00°$ C. |
| The melting point of ice ... | ... | $0·00°$ C. |
| The boiling point of oxygen | ... | $-182·97°$ C. |

As secondary fixed points the melting point of mercury $(-38·9°$ C.) and the sublimation point of carbon dioxide $(-78·5°$ C). are recommended.

# CHAPTER V

## RECTIFICATION IN THEORY
## AND PRACTICE

### I. v. 1. *Binary Gases and their Equilibrium with Liquids*

By far the most important industrial application of Low Temperature Physics is the production of pure gases. It may appear strange that physical methods should have prevailed in this domain of applied science, for from the point of view of the phase rule a gaseous mixture is inseparable by physical means. It is therefore not surprising that the physical methods are characterised by the fact that at least one of the gases is liquefied in the course of the process. This is indeed the reason why low temperatures are employed in the separation of gaseous mixtures.

As long as the boiling points of the gases in question lie very far apart, separation can be effected simply enough by liquefying one component and cooling so far that the partial pressure of this component is negligible This method is in fact employed for removing benzene from coke gas and in several other cases. It is when the boiling points of the components lie so close together that it is impossible to liquefy one without the other that the difficulties begin. This problem was solved in principle when Linde adapted the process known as rectification in the alcohol industry to low-boiling mixtures.

The thermodynamical theory of systems containing more than one component, a theory developed by van der Waals and his collaborators, is an exceedingly intricate matter. We shall here content ourselves with a simplified summary, giving only a few of the characteristic lines of thought and outlining the geometrical methods employed. We shall discuss only systems containing two components, for which the classical example is oxygen and nitrogen.

We may obtain a good survey of what to expect from a theory of binary mixtures by applying the phase rule in the form $k - \phi = f - 2$, where $k$ signifies the number of components,

$\phi$ the number of phases and $f$ the number of degrees of freedom. A binary gaseous mixture in equilibrium with its liquid gives $k = 2$, $\phi = 2$, and therefore $f = 2$. Now such a system possesses two variables, apart from temperature and pressure, namely the concentrations $c_A^l$ and $c_A^g$ of the component $A$ in both phases. The concentrations of the other component $B$ are then of course $1 - c_A^l$ and $1 - c_A^g$. The phase rule thus states that we are at liberty to fix two of the variables within certain limits, whereupon the other two are automatically determined.

Let us assume for the moment that the pressure $p$ and $c_A^g$ are fixed and that the temperature $T$ and $c_A^l$ are to be determined. Consider 1 gramme-molecule of a mixture of oxygen and nitrogen gas containing $n$ per cent. nitrogen to be cooled at atmospheric pressure. At what temperature does liquefaction begin and what is the composition of the liquid phase?

The answer to this is given in fig. 1, which shows the equilibrium curves of an oxygen-nitrogen mixture at 1 atm. pressure. The concentrations are plotted as abscissae, pure nitrogen on the left and pure oxygen on the right, and the ordinates are temperatures. The curves were first measured by Bailey and later repeated with greater accuracy by Dodge and Dunbar. The upper curve is known as the dew curve, the lower as the boiling curve. Above both curves we have only one phase, and that is gaseous; below both curves the substance is entirely liquid. Only on and between the two curves does equilibrium exist between the two phases.

In fig. 1 horizontal lines mark isothermals and vertical lines are lines of constant composition. In order that phases may be in equilibrium with one another they must have equal temperature. Thus to every point on one curve corresponds one and only one point on the other curve situated on the same horizontal. States denoted by these two points are in equilibrium with one another.

Now suppose the concentration we have chosen to be denoted by the vertical dotted line $AA'$. When we move downwards along this dotted line liquefaction begins at point $P$ and the composition of the liquid in equilibrium with our gas at this temperature is given by point $Q$ on the boiling curve. Thus the first drop of liquid that is formed when we reach $P$ has the composition $Q$ and therefore contains considerably more oxygen than the gas. If the temperature is kept constant, no more liquid is formed; in order to allow liquefaction to proceed, the temperature must be lowered further. The composition of both phases now changes, that of the gas moving along the

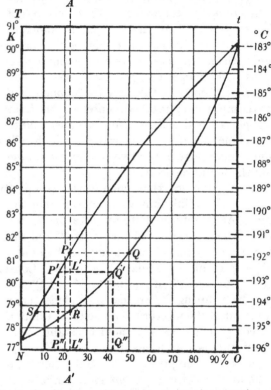

Fig. 1. Gas liquid equilibrium in oxygen nitrogen mixtures.

dew curve to the left and that of the liquid along the boiling curve in the same direction. Finally, when all the gas is liquefied, the composition of the liquid must obviously be the same as the original composition of the gas before liquefaction began. The end of condensation is thus given by point $R$ on the boiling curve, that of the last bubble of gas being shown by $S$ on the dew curve. We see that in all stages of the process of condensation the liquid contains more oxygen than the gas.

5-2

This is one of the two leading facts on which rectification is based. The other becomes apparent when we consider two points, one on each curve, situated vertically one below the other, i.e. on a line of constant composition, for instance, $P$ and $R$. The states denoted by these two points, being at different temperatures, are not in equilibrium, and this is evidently true for all such points on the same vertical line except at the ends of the diagram, where the two points coincide. Thus a gas and a liquid having the same composition are never at equilibrium with one another. If we bring them together, their compositions will change, the gas giving off one component to the liquid or absorbing the other component out of the liquid, until equilibrium is reached. Moreover, the temperature of the two phases will change simultaneously, until finally both will assume the same temperature, which will be the equilibrium temperature of the two compositions assumed by the gas and the liquid respectively.

Fig. 1 shows us more than merely the compositions of liquid and gas at equilibrium with one another at various temperatures. At an intermediate temperature $L'$ between $P$ and $R$, when the composition of the gas is $P'$ and that of the liquid $Q'$, we may calculate the amount of liquid and the amount of gas present in the mixture as follows. Consider the projection $P''$, $L''$ and $Q''$ of points $P'$, $L'$ and $Q'$ on the axis of abscissae. Now during the process of condensation not only the total amount of the mixture, but the total amount of each one of the components, must remain constant. If the total amount of mixture is 1 gramme-molecule, $OL''/100$ gives the total amount of nitrogen present. If $l$ be the amount of liquid and $1-l$ the amount of gas present at a temperature corresponding to $L'$, then $lOQ''/100$ is the amount of nitrogen in the liquid and $OP''/100 . (1-l)$ the amount of nitrogen in the gas.

Therefore
$$\frac{OQ''}{100}l + \frac{OP''}{100} - \frac{OP''}{100}l = \frac{OL''}{100},$$

or
$$l = \frac{OL'' - OP''}{OQ'' - OP''} = \frac{P''L''}{P''Q''} = \frac{P'L'}{P'Q'},$$

and similarly
$$1 - l = \frac{Q''L''}{P''Q''} = \frac{Q'L'}{P'Q'}.$$

Thus the ratio of liquid to gas is given by $P'L' : Q'L'$, a formula that is valid for all temperatures between $P$ and $R$. At point $P$ itself the amount of liquid is zero, and similarly there is no gas at $R$. The rule enabling us to compute the amount present in both phases from points on the diagram is known as the "lever rule" and is of great importance in the theory of rectification.

The general shape of the curves in fig. 1 appears in a number of binary mixtures and is typical for the simple case when no chemical compounds are formed and the liquid components are completely miscible in all concentrations.

### I. v. 2. *The Rectification Column*

The simplest type of rectification plant consists of two parts, the column $C$ and the evaporator $E$, as demonstrated in fig. 2. The former is a heavily insulated vertical cylindrical vessel, in which a number of perforated metal pans $P_1$, $P_2$, ... are fixed horizontally at definite intervals. The evaporator is simply a metal boiler fitted out with an appliance for heating the liquefied gas, which we shall assume to be a mixture of nitrogen and oxygen.

Suppose a certain quantity of liquid air to be poured in from above at $L$ in the unit of time. The liquid will possess approximately the composition of atmospheric air, viz. 79 per cent. $N_2$ and 21 per cent. $O_2$. It will collect in the pans and gradually trickle down into $E$, where we will assume that it is completely evaporated. The gas formed, which will thus have about the same composition as the liquid, will then bubble up through the pans, where it will mix with the liquid phase.

According to the laws developed in the last paragraph, the gas arriving at the lowest pan will give off oxygen to and absorb nitrogen from the liquid air in the pan until approximate equilibrium is reached. This process will be continued in the next pan, and so on. The liquid now leaving the lowest pan in a downward direction will thus contain more oxygen than the first portion, the same applying to the next portion of gas leaving the evaporator. It is thus clear that in course of time the liquid entering and the gas leaving the evaporator will come to be pure oxygen. A certain portion of the liquid oxygen may be tapped off through $R$, the rest being needed to carry on the process of rectification. It is clear that this type of apparatus, the so-called single-column plant, can produce pure oxygen but not pure nitrogen. For the gas leaving the top of the column at $G$ can at the best contain no greater percentage of nitrogen than is in equilibrium with the liquid air entering at $L$. A glance at Bailey's curves in fig. 1 will show that the gas in equilibrium with a liquid mixture containing 79 per cent. of nitrogen will itself contain 93 per cent. of nitrogen. Therefore not only will the nitrogen emerging at $G$ be impure, but 28 per cent. of the oxygen employed is lost in the course of the process. Nevertheless, the single-column plant is still frequently employed in order to obtain pure oxygen, the loss of oxygen being compensated by the cheapness and simplicity of the apparatus.

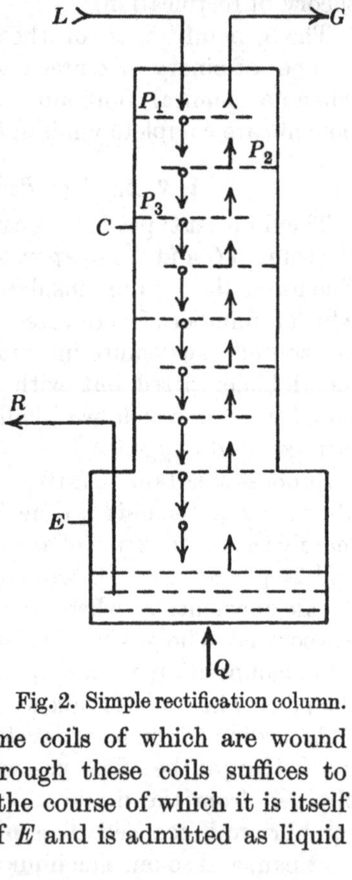

It is obviously not practical to heat the evaporator with an electric furnace in order to evaporate the oxygen. In practice the air is admitted

Fig. 2. Simple rectification column.

through a heat exchanger, some coils of which are wound inside $E$. The gas passing through these coils suffices to evaporate the oxygen in $E$, in the course of which it is itself liquefied. It then passes out of $E$ and is admitted as liquid at $L$ as shown in the figure.

In order to enable pure nitrogen as well as pure oxygen to be produced in the column, it is necessary to prolong the latter above the air-inlet and simultaneously to cool the out-going gas and to condense part of it at the temperature of boiling nitrogen, which, as fig. 1 shows, is considerably lower than that of liquid air. In this way the purification process can be continued above $L$ and pure nitrogen be gained in the condenser. However, in order to cool to the temperature of

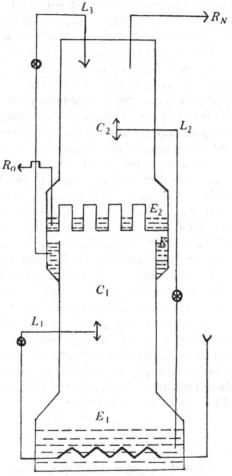

Fig. 3. Double air rectifier.

boiling nitrogen, we must presuppose the existence of pure nitrogen as a cooling agent, whereas the production of this pure nitrogen is the aim of our process. To overcome this difficulty Linde introduced the two-column rectifier, a sketch of which is shown in fig. 3.

Instead of cooling the condenser with liquid nitrogen Linde raises the pressure in the entire column $C_1$ to about 5 atm. In this way the boiling point of nitrogen is raised above that of oxygen at normal pressure, and it is thus possible to cool the condenser with liquid air or liquid oxygen. This is effected as follows: the column $C_1$, working at 5 atm., effects only a preliminary purification of the mixture. The pure gases are gained in the upper column $C_2$ at atmospheric pressure, the condenser $K$ of $C_1$ being combined with the evaporator $E_2$ of $C_2$. The liquid air is admitted under pressure in the middle of $C_1$ at $L_1$. It is not allowed to form pure oxygen in $E_1$, but a mixture containing only about 40 per cent. of oxygen. This is brought about by tapping off more liquid from $E_1$ than would be permissible for the production of pure oxygen. However, the nitrogen condensed in $K$ at the top of $C_1$ can be made as pure as we wish. The liquid mixture leaving $E_1$ is expanded to atmospheric pressure and admitted at $L_2$ in the middle of $C_2$, and the pure liquid nitrogen in $K$ is partly removed and admitted at $L_3$ to the top of $C_2$. Pure oxygen is then formed in $E_2$ and can be removed through $R_O$, and pure gaseous nitrogen can be removed at $R_N$. If, as is usually the case, the aim of the process is to produce oxygen and nitrogen in the gaseous state and at room temperature, the cold gases tapped off are passed out through heat exchangers, thereby cooling down the in-going air. By regulating the amount of liquid removed from $E_1$ and $K$ to the upper column, the degree of purity of the resulting gases can be adjusted to meet the requirements of industry. Usually a very high purity of one gas entails slight impurities in the other.

### I. v. 3. *"Rectification Calculus"*

It will be evident to anyone carefully studying fig. 3 that we have given but a bare outline of the facts. To even a casual observer a host of questions must immediately arise. What is meant by "pure" nitrogen? What degree of purity can be reached? Why is a 40 per cent. oxygen mixture chosen in $E_1$ and how much is tapped off to $L_2$? How much liquid nitrogen is admitted at $L_3$ and how much flows back through $C_1$? How

many pans are needed to effect a certain degree of rectification ?
etc. etc.

All these questions can be answered and have been answered, not only as the result of laborious experiments carried out in the apparatus itself, but also theoretically with the help of a very simple calculus first suggested by Keesom and developed and adopted to the problem by Weissberg.

Though it would take us too long to give anything like a full computation of a rectifying column, the method is so elegant

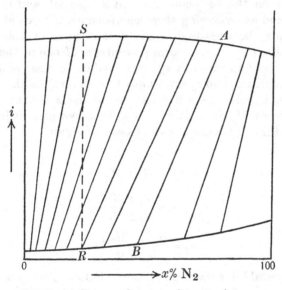

Fig. 4. $i$-$x$ diagram of oxygen-nitrogen mixtures.

and so little known to the general physicist that a short outline appears appropriate.

The basis of operations is the $i$-$x$ diagram of oxygen-nitrogen mixtures shown in fig. 4 for atmospheric pressure, in which the abscissae are concentrations measured in percentages of $N_2$ and the ordinates total heat, which may be measured in calories per gramme-molecule or per any other unit of mass or volume. The curves, which mark phase equilibrium, are constructed from specific heat data, the lines join corresponding points on the dew curve and boiling curve and are derived from fig. 1. Thus a saturated gas, whose state is given by point $A$, is at equilibrium with a liquid at point $B$, and so on. These lines are known as connodals, an expression derived from Korteweg's theory of surfaces. As $A$ and $B$ are in equilibrium with each other, they possess the same temperature, and the connodals thus correspond to the horizontal lines in fig. 1.

Now in order to move about freely on this diagram Keesom and Weissberg have developed the following calculus. A point $A$ on the $i$-$x$ plane is called a "phase" and is to be considered as possessing three components, a "weight" $M_A$, which may be measured in any unit of mass or volume, a composition $x_A$, measured in percentages of one of the components, and an enthalpy $i_A$, measured in calories per unit of $M_A$. Whereas $x_A$ and $i_A$ can simply be read off from the figure, $M_A$ cannot. This drawback is lessened by the fact that in many calculations $M_A$ eventually cancels out. We now define the sum and the difference of two phases as follows:

$$C = A \pm B,$$

when
$$M_C = M_A \pm M_B,$$

(I)
$$x_C = \frac{M_A x_A \pm M_B x_B}{M_A \pm M_B},$$

$$i_C = \frac{M_A i_A \pm M_B i_B}{M_A \pm M_B}.$$

The physical interpretation of this definition is easily discernible (see fig. 5). Suppose we bring two gaseous "phases" together, $M_A$ litres of gas given by point $A$, and $M_B$ litres of a mixture given by point $B$. The result will obviously be $C$ litres of a gas given by point $C$, the "weight" of the resultant mixture, i.e. the total amount of gas, being given by the sum of the weights of the two component mixtures, and composition and enthalpy being found with the help of the rule given above. It will be remembered that a similar rule was derived in the last paragraph to determine the relative amounts of liquid and gas on the $T$-$x$ diagram.

A similar interpretation is apparent for the "difference" of two phases. If from a phase $A$ we remove a phase $B$, the result is a phase $C$, the properties of which are in fact given by our definition of the difference $A - B$. We may go slightly farther in our interpretation of these rules. Suppose we have a gaseous mixture $A$ in equilibrium with a liquid mixture $B$ as in fig. 4. We may then combine these two phases by considering them as equivalent to a single phase $C = A + B$ situated on the connodal $AB$ at a distance from the points $A$ and $B$ as given by rule (I). Physically we may interpret this as follows. Suppose a certain binary mixture to have been brought by some means or other to the state denoted by $C$ between the dew line and the boiling line of our $i$-$x$ plane. Then, as we

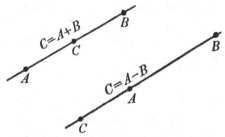

Fig. 5. Addition and subtraction of phases.

know, the mixture will not remain in $C$ but will promptly divide into a liquid phase $B$ and a gaseous phase $A$ situated at the ends of the connodal through $C$, the weights being determined by the distances from $C$ to $A$ and $B$ on the connodal. However, we must note that our calculus alone is naturally unable to determine $A$ and $B$. On every straight line through $C$ any point $A'$ will determine a corresponding point $B'$ such that $A' + B' = C$. To fix $A$ and $B$ three mathematical data are needed, the direction of the straight line through $C$, the distance of one of the points $A$ and $B$ from $C$ and either the position of the other point or the weight of the first. Physically these three data are given by the two equilibrium curves and the direction of the connodal.

One more definition is needed to make our calculus complete. Suppose, for instance, that we have a liquid mixture on the boiling curve given by a point $R$ (fig. 4). By applying a quantity of heat equal to the total heat of evaporation we may bring the mixture to point $S$ on the dew curve. The resultant gas will then of course have the same weight and the same composition as the original liquid. The only difference between the two phases is that $S$ has the greater enthalpy. We shall therefore find it convenient to define the sum and difference of a phase $A$ and a quantity of heat $Q$ as follows:

$$B = A \pm Q$$

when $\qquad M_B = M_A, \quad x_B = x_A, \quad i_B = i_A \pm Q/M_A.$

$Q$ is then measured in calories. We may leave it to the reader to check the commutative and associative rules contained in our calculus and may with this equipment proceed to sketch the method employed to calculate a rectifying column.

For the sake of simplicity we shall consider a column of the simple type shown in fig. 2, and suppose that a stationary state has been reached. Then it is evident that in every section of the column the amount of substance present, its composition and its total heat are independent of time. Now consider a certain section containing one pan, say pan No. 1 (fig. 6). Since the state is stationary, amount, composition and total heat of the mixture leaving the section are equal to those of the mixture entering. Now a gas $G_0$ enters from below and a liquid $L_2$ from above, whereas a gas $G_1$ leaves the top and a liquid $L_1$ the bottom of the section. Thus according to our calculus

$$G_0 + L_2 = G_1 + L_1 \quad \text{or} \quad G_0 - L_1 = G_1 - L_2.$$

$G_0 - L_1$ is the "phase" entering the section from below and $G_1 - L_2$ the "phase" leaving it from above. But $G_1 - L_2$ is similarly the phase entering the next section from below, and it may similarly be shown that

$$G_1 - L_2 = G_2 - L_3$$

and in general that

$$G_{n-1} - L_n = G_n - L_{n+1} = Z.$$

We have thus shown that *the "phase" moving through the column in a given direction is constant throughout the column.*

We shall now assume that no interaction takes place between gas and liquid between two pans, but that on the pans themselves complete equilibrium is reached between the two phases.

This means that whereas gas and liquid mixtures passing each other between two pans are not in equilibrium, the gas leaving the top of a pan is in equilibrium with the liquid leaving the bottom.

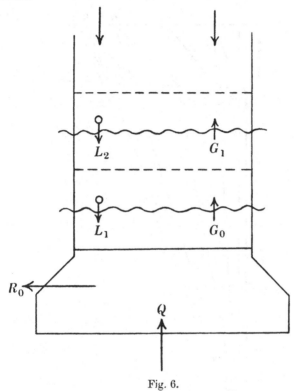

Fig. 6.

Now let us turn to fig. 7. Suppose that we introduce at the top of the column a certain quantity of liquid air per minute containing 79 per cent. of nitrogen. The phase of this liquid is given by point $A$. Suppose we wish to remove from the top of the column a certain quantity of gas containing 91 per cent. of nitrogen, i.e. a gas not quite so pure as that in equilibrium with $A$. The phase point corresponding to this gas is $B$. Since we know the weights of the gases, we may calculate the difference $A - B = Z$. Now the gas $B$ leaving the column at the top is in equilibrium with the liquid leaving the uppermost pan in a downward direction. The phase point of this liquid may therefore be found by drawing the connodal $BC$. Now we have shown that the phase point of the gas passing $C$ between the uppermost pan and the next must be

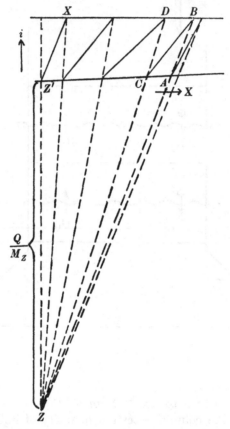

Fig. 7. Determination of number of pans.

situated at a point $D$ of the dew curve on the straight line $CZ$. For $A - B = C - D = Z$. In this way we may find all the other phase points of the column.

How far can the process of rectification proceed? Obviously not *ad infinitum*! By fixing the weights and compositions of the phases entering and leaving the top of the column we have indeed already determined the degree of purity of the oxygen and the number of pans needed. For the liquid phase tapped off from the evaporator may be calculated for the stationary state. If the liquid phase $L_1$ enters the evaporator and the gas $G_0$ leaves it, and if we remove a liquid phase $R_0$ and simultaneously apply a quantity of heat $Q$, then

$$L_1 + Q = G_0 + R_0$$

or $$R_0 = L_1 - G_0 + Q = Z + Q = Z'.$$

Weight and composition of the mixture produced are thus given by the known components of $Z$. Moreover, the heat required in the evaporator is also determined; for the liquid that we remove comes straight from the evaporator and is therefore in equilibrium with a vapour of some composition or other. $Z'$ must therefore be on the boiling curve. $Q/M_Z$ is therefore given by the vertical distance from $Z$ to the boiling curve.

In order to determine the number of pans needed under the given conditions we have only to repeat the process described above until our phase points have moved across the figure from $AB$ to $Z'X$. The number of steps taken gives the number of pans required.

In the above example we have fixed arbitrarily the amount and composition of the liquid entering and of the gas leaving the column at the top. The first two of these magnitudes are generally given by the atmospheric air at our disposal and the dimensions of the plant. Instead of the second two it is usually more to the point to lay down the composition of the "oxygen" removed from $E$, and the amount of heat $Q$ evolved in the evaporator per unit of $R_0$. This fixes the points $A$ and $Z'$ in fig. 7 as well as $Z$, since the vertical distance $ZZ' = Q/M_{Z'} = Q/R_0$. Point $B$ is then obtained by connecting $Z$ and $A$ and producing to the dew curve. The weights of $B$ and $Z$, and therefore of $Z' = R_0$, may then be found by applying the lever rule along the line $ZAB$, the number of pans being computed as before. The problem may in fact be solved whenever any two data are given apart from $M_A$ and $x_A$. When $A$ and $x_{Z'}$ are given it is evident that a minimum value exists for $Q$. For the point $Z$ must be situated low enough that the extension of $ZA$ may

intersect the dew curve to the left of the connodal through $A^{\cdot}$
Moreover, if $Z'$ lies very far to the left, where the connodals are
almost vertical, very many pans are required near the bottom
of the column. This is equivalent to saying that very pure
oxygen necessitates very many pans. The less stringent the
requirements on the purity of the gas, the fewer pans are
needed.

This example may suffice to show how the method here
described enables us to calculate the various parts and properties of a rectifier. An application of the method to the two-
column apparatus, which would supply us with answers to all
the questions mentioned above and to a great many more,
would however take us beyond the scope of this book.

The energy required in theory to separate 1 kg. of air into
its two principal components may be computed very simply
without any special calculus with the help of Gibbs' formula
for the entropy of a perfect gaseous mixture:

$$S_m = \sum_{\nu} n_{\nu} \left( C_{p\nu} \log T - R \log p_{\nu} + k_{\nu} \right),$$

where $n_{\nu}$ is the number of gramme-molecules of the $\nu$th component, $p_{\nu}$ the partial pressure, $k_{\nu}$ a characteristic constant of
the gas and $C_p$ and $R$ have the usual significance. Since the
energy of a perfect gas depends on temperature alone, the
minimum quantity of work needed to separate the mixture at
a definite pressure $p$ into components each at this same
pressure $p$ is given by

$$A_{\text{min.}} = T\Delta S = T \left\{ \sum n_{\nu} \left( C_{p\nu} \log T - R \log p_{\nu} + k_{\nu} \right) - \sum n_{\nu} \left( C_{p\nu} \log T - R \log p + k_{\nu} \right) \right\},$$

the last expression on the right-hand side expressing the
entropy of all the gases separately. Therefore

$$A_{\text{min.}} = RT \sum n_{\nu} \log \frac{p}{p_{\nu}},$$

or since $\qquad n_{\nu} = \frac{n}{p} p_{\nu}$ with $\sum n_{\nu} = n,$

$$A_{\text{min.}} = nRT \sum \frac{p_{\nu}}{p} \log \frac{p}{p_{\nu}}.$$

This is exactly the amount of work needed to compress each gaseous component separately from its partial pressure to the total pressure of the mixture. If we apply this formula to an oxygen-nitrogen mixture of atmospheric composition at room temperature and atmospheric pressure, neglecting argon and the other rare components, we obtain $A_{min.} = 0 \cdot 0116$ kw.h. as the minimum quantity of work required to separate 1 kg. of mixture into the oxygen and nitrogen, each again at a pressure of 1 atm. This is very much less than the work needed to liquefy the same quantity of air.

Hausen has described an "isentropic rectifier" with which separation could in principle be effected with this minimum quantity of work. As yet the practical efficiency of rectifiers is, however, small and 15 per cent. is considered quite a good figure. A large column requires about $0 \cdot 08$ kw.h. to separate 1 kg. of air into almost pure oxygen and nitrogen.

## I. v. 4. *The Production of Pure Gases*

Nature has made the most of the fact that the entropy of gases increases on mixing. Of all the natural gases not one occurs in an undiluted condition, and the same must be said of almost all that are obtained in the course of technological processes. We have now nearly forgotten that the purification of gases is a very serious problem and that before the beginning of the present century pure gases were practically unknown outside the laboratory. Now, though in some domains chemical methods prevail, the separation of gases is the one great sphere of applied cryogenics. The following gases are obtained either exclusively or at least to a great extent with the help of low temperatures: nitrogen, oxygen, hydrogen, helium, argon, neon, krypton and xenon. Recently methane was added to the list and apparently ethylene will shortly follow. Three primary sources serve for the production of these gases. The first is *air*, which yields oxygen and nitrogen and all the rare gases with the exception of helium. The latter is now obtained from the second source, which is usually termed *earth gas*. This is a general expression for several natural gases which the earth emits in a number of places and which contain methane, nitrogen, occasionally $CO_2$ and helium. The percentage of helium varies from some hundredths of a per cent. to 2 per cent., the gases richest in helium occurring in North America. The third source is *coke gas*, which is obtained in the course of converting coal to coke. This gas contains from 45 to 60 per cent. of hydrogen, for which it is one of the principal sources. The chief component of coke gas besides hydrogen is methane, the technical application of which is only just commencing.

Pure *nitrogen* and *oxygen* are obtained from air in the double rectification plant the principle of which we have treated. Several types of these plants are in use, as Linde, Claude and Heylandt have developed their liquefaction principles for use in rectification. The separation of air was indeed the main purpose which led these men to construct their liquefiers. The Claude and Heylandt types are most suitable for producing the pure gases in the liquid form; for obtaining gases at room temperature Linde's construction has numerous advantages. In these plants 80 per cent. of the air is admitted at 5 atm. as required by the lower column, and the remaining 20 per cent. compressed to 200 atm. and expanded in a throttle valve, to maintain the low temperatures; for the rectifying plant is a refrigerating machine in the same sense as the liquefier, even when the products are obtained at room temperature.

Fig. 8 shows a photograph of a small Heylandt rectifier in the low-temperature laboratory at Kharkov, which gives 25 litres of liquid nitrogen an hour. The column is about 18 ft. high and works at 180 atm. 60 per cent. of the air enters the expansion cylinder at room temperature, where it is cooled to $-125°$ C. The same model may be seen in the Royal Society Mond Laboratory at Cambridge.

In recent years *argon* has come into demand for filling incandescent lamps. The argon content of air is 0·93 per cent. by volume. The concentration of argon is effected by tapping off a mixture from the air-rectifying column at a point where its concentration is highest and rectifying this mixture in a separate column cooled by liquid nitrogen. From empirical data which confirm the calculations of Fischer and Hausen,

Fig. 8.

Heylandt rectification plant.

the argon concentration is highest slightly above the evaporator of the upper column and amounts to about 4 per cent.

Liquid is therefore withdrawn from the air column at this point (1) and introduced into the argon column at (2), as shown in fig. 9. Gaseous nitrogen is removed from the air-column condenser at (3), condensed in the bottom of the argon column at (4), thus evaporating the argon fraction, expanded to atmospheric pressure in the valve $V$ and introduced at the top of the column at (5), where it serves to condense the argon. The concentrated argon fraction emerges at (6) and pure oxygen at (7). In the argon column, which is generally placed beside the air column and is insulated in the same casing, it is not possible to produce pure argon. Generally 50 per cent. of argon is obtained, the rest being mainly oxygen. This is removed chemically with hydrogen or sulphur, leaving a mixture known as technical argon containing about 12 per cent. of nitrogen. This is about the composition which is required by the incandescent lamp industry.

Hausen recently showed that argon plays a more important part in air rectification than might have been expected from its small percentage. In fact the lower part of the upper air column contains considerably more argon than nitrogen. Hausen pointed out that in order to obtain sufficiently pure oxygen it is indispensable to remove the argon from the column, as the presence of argon very definitely impedes the final stages of rectification.

In large air plants it was found advisable to blow out nitrogen from the condenser occasionally in order to maintain the full efficiency. This was discovered to be due to a cushion of *neon-helium* mixture, which is formed at the top of the lower column. Now that small quantities of neon are required for lamps and advertisements, it is customary to purify this mixture further. The gas blown off consists of about 70 per cent. $N_2$, 23 per cent. Ne and 7 per cent. He. By compression in liquid air the mixture can be freed from nitrogen. The helium is usually not removed. Separation of neon and helium can be accomplished only at hydrogen temperatures, at which neon solidifies, yielding pure helium. By this method Meissner secured sufficient helium to work a helium liquefier,

6-2

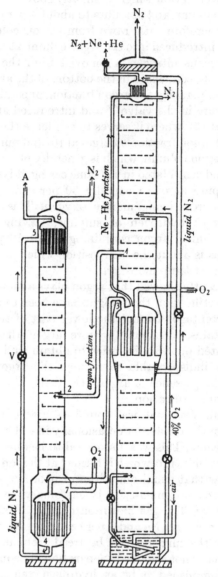

Fig. 9. Rectification plant with argon column.

but it took him several years to collect. Atmospheric air contains

$$1\cdot5 . 10^{-3} \text{ per cent. by volume Ne,}$$

$$5 . 10^{-4} \text{ per cent. by volume He.}$$

The production of *krypton* and *xenon* on a large scale has not yet been possible owing to the minute quantities contained in air. The figures are

$$\text{Kr } 1\cdot08 \pm 0\cdot10 . 10^{-4} \text{ per cent. by volume,}$$

$$\text{X } 0\cdot08 \pm 0\cdot03 . 10^{-4} \text{ per cent. by volume.}$$

These gases may be obtained in small quantities from the evaporators of air-rectifying plants, where they gradually collect, but even here the concentration remains small. Several methods have been employed to concentrate them, the most promising of which appears to be one recently suggested by Claude, in which large quantities of the gas are washed in small quantities of liquid air. As a krypton-xenon mixture raises the efficiency of incandescent lamps some 30 per cent. as compared with argon, it may be worth while attempting to produce these gases in large quantities.

The separation of krypton-xenon mixtures is still in the laboratory stage. For the small quantities hitherto obtainable the most suitable method is fractional desorption, as suggested and carried out by Peters and Weyl. The method is based on the fact that if an equal amount of krypton or xenon is adsorbed on silicagel or charcoal, their adsorption pressures at a given temperature differ more than their vapour pressures at corresponding temperatures. Fractional desorption thus leads to a more efficient separation than fractional distillation.

Pure *helium* may be obtained from earth gas by fractional condensation, as its boiling point is very much lower than that of any of the other components. Separation is therefore comparatively simple so long as the helium concentration is not too small, in which case the solubility of helium in the liquid phase leads to very material losses. For very weak helium sources a suitable method has not yet been developed.

Before leaving the rare gases we may sum up a few of their characteristic data:

### Table VI

| Gas | $V$ per cent. | $T_b$ | $T_t$ | $p_t$ (cm.) |
|-----|---------------|-------|-------|-------------|
| He  | $5.10^{-4}$   | 4·19  | .     | .           |
| Ne  | $1\cdot5.10^{-3}$ | 27·17 | 24·57 | 32·35   |
| A   | 0·93          | 87·25 | 83·78 | 51·87       |
| K   | $1\cdot1.10^{-4}$ | 120·86 | 116·2 | 52·22   |
| X   | $0\cdot8.10^{-5}$ | 164·5 | 161·2 | 61·55   |

In Table VI $V$ per cent. signifies the percentage by volume in atmospheric air, $T_b$ and $T_t$ the boiling point and triple point respectively and $p_t$ the triple-point pressure in centimetres of mercury. It is of interest to note the high values of $p_t$, increasing with rising atomic weight, and the exceptional position of helium, which has no triple point.

The purification of *hydrogen* from coke gas is by far the most complicated process hitherto attempted by low-temperature engineering. The increasing demand for pure hydrogen and for hydrogen-nitrogen mixtures in the production of synthetic ammonia has led to the construction of gigantic plants, although this is only one of the several methods for obtaining hydrogen.

The composition of coke gas varies with the nature of the coal and the details of the treatment. Table VII gives a typical analysis:

### Table VII

| Gas | $V$ per cent. | | $V$ per cent. |
|-----|---------------|---|---------------|
| $H_2$   | 57·7 | $O_2$      | 0·9 |
| $CH_4$  | 24·5 | $C_2H_4$   | 1·8 |
| $N_2$,  | 8·7  | $C_2H_6$   | 0·9 |
| $CO$    | 5·5  |            |     |

Analysis after the removal of benzene and naphthalene. The figure for $C_2H_4$ contains a certain amount of higher unsaturated hydrocarbons.

The separation is effected, according to Linde, in an intricate apparatus in which the gas, at a pressure of 13 atm., is successively cooled in a system of heat exchangers and finally treated with liquid nitrogen. In the course of this process three fractions are obtained, known as the ethylene, methane and carbon monoxide fractions. The first contains mainly ethylene and methane, the second, apart from methane, oxygen, carbon monoxide and nitrogen, and the third mainly the last two gases. The temperature in the apparatus varies from $-45°$ to $-192°$ C. from point to point. The entire process is based on liquid nitrogen, which serves three separate purposes. First, it supplies all the cold required to maintain the low temperatures. Secondly, it is used to wash out all the carbon monoxide by rectification, and thirdly, a definite quantity is added to the final product to produce the stoechometrical mixture for the synthesis of ammonia. The removal of the last traces of carbon monoxide is essential, as the latter poisons the catalysts.

Thus each aggregate is accompanied by a large air-rectifying plant, which produces the necessary quantity of nitrogen. This is an evident drawback, as it complicates the process: air must be compressed and liquefied, whereupon, after rectification, nitrogen is obtained as a gas at room temperature and atmospheric pressure. It must then be again compressed to 200 atm. and liquefied afresh. Though this process is carried out on a gigantic scale it is still a young and incompletely developed industry and is probably not the last word in the production of hydrogen with the help of low temperatures. The method employed by Claude is rather simpler but yields hydrogen of lower purity.

# CHAPTER VI

## SOLID LIQUID EQUILIBRIUM

### I. VI. 1.   *The Equilibrium Curve*

In fig. 1 is shown the usual form of the phase diagram of a one-component system. Pressure is plotted as ordinates and temperature as abscissae. The curves mark equilibrium between two phases. Beginning at absolute zero at a pressure which is either itself zero or at any rate very small, the *sublimation curve a* denotes the boundary of the solid and gaseous

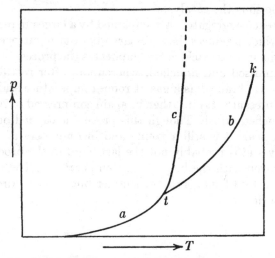

Fig. 1. Equilibrium curves of one-component system.

states. It rises gradually to the *triple point t*, at which the solid body melts in equilibrium with its gaseous phase. With very few exceptions the triple-point pressure $p_t$ is less than 1 atm. From this point two fresh curves originate, the *vapour-pressure curve b*, marking the boundary between liquid and gas, and the *melting curve c*, which divides liquid from solid. The vapour-pressure curve is well known for most substances and we know that it terminates in the *critical point k*, above which the liquid and gaseous phases are identical.

The melting curve *c*, which we shall treat in this chapter, is still a problem, and until quite recently all that was known of it was that it rose very steeply, i.e. that very considerable pressures are needed to produce a notable change in the melting point of a solid. In connection with this curve two questions are important: (1) What is the general shape of the curve? (2) Does it lead to a critical point, i.e. does a limiting temperature exist above which a crystal and a highly compressed gas or liquid may no longer be distinguished?

From the experimental side solid-liquid equilibrium was first investigated systematically by Tammann, who measured the melting curves of a number of organic substances and some low-melting metals up to 300 atm., as well as several more complicated substances such as double salts. Tammann expressed his results in the following form:

$$T_m = T_l + ap_m - bp_m^2,$$

for
$$p_m \geqq p_l,$$

where $T_m$ and $p_m$ are respectively the temperature and pressure of melting and $a$ and $b$ are positive constants. It is clear that according to Tammann's formula the melting curve not only grows steeper as the temperature rises but becomes retrograde after a certain temperature, a rise of pressure lowering the melting temperature. This would mean that the crystal state occupies a closed area on the $p$-$T$ diagram, and that at a certain finite pressure $p_0$ solid and liquid are in equilibrium at absolute zero. Tammann, who came to this conclusion as a result of experiments on double salts, upheld his opinion in spite of the fact that equilibrium between solid and liquid at zero temperature and under finite pressure is hard to reconcile with Nernst's theorem (see Part II, chap. III).

Some years later P. W. Bridgman, who has developed high-pressure technique to an astounding degree of perfection, began measuring melting curves up to several thousand atmospheres. Though Bridgman's work has every right to be termed classical from the point of view of accuracy and thoroughness, his results were not very encouraging. The melting curve continued to grow steeper as the pressure was raised, yet retrograde curves were not found and Tammann's simple formula was not corroborated; neither were any signs of approaching critical phenomena apparent.

It was at this point that Simon approached the problem with the methods of low-temperature physics. Simon argued roughly as follows: in order to discover the ultimate shape of the melting curve it is evidently necessary to follow up this curve farther than either Tammann or Bridgman were able to do. On the other hand there seems to be no possibility of working at higher pressures than Bridgman's, for the simple reason that no material exists that will stand such pressures. However, a way to the solution is laid down by the principle of corresponding states. What from the point of view of this principle can be meant by following up the curve to "high temperatures"? High and low temperatures can have significance only in relation to the substance under investigation. We are therefore concerned primarily not with the production of temperatures and pressures that are absolutely high but with pressures that will bring the melting point of our substance to temperatures which are high as compared with the normal melting point of the substance. As we cannot vary our pressure to meet this demand, our only possibility is to vary the melting point, i.e. the substance under investigation. The question is—what type of substance is most suitable? This question may be answered by considering a formula which Simon developed more or less empirically and which covered Tammann's and Bridgman's results very satisfactorily:

$$\log(p+a) = c\log T + b. \qquad \dots\dots(1)$$

Here $p$ is the pressure at which the substance melts at the temperature $T$ and $a$, $b$ and $c$ are constants depending only on the substance. A short discussion of this formula will be useful, especially as regards the physical interpretation of the constants. The constant $b$ may be easily accounted for with the help of the triple point $T_l$ if we note that the pressures considered are of the order of several hundreds or even thousands of atmospheres. The triple-point pressure $p_l$, which is usually some tenths of an atmosphere, may thus be neglected, and by inserting $T_t$ in (1) we obtain

$$\log a = c \log T_t + b,$$

whence
$$\log \frac{p+a}{a} = c \log \frac{T}{T_t} = c \log T'. \qquad \ldots\ldots(2)$$

Having thus eliminated $b$ we see that it is profitable to introduce $T/T_t = T'$ as a reduced co-ordinate, the triple point $T_t$ taking the place of the critical point $T_c$ in a new "law of corresponding states".

The constant $a$ may be interpreted by considering that the pressure 0 is not significant for a solid body; for apart from the pressure that may be exerted upon the body from without, every solid body is subjected to a powerful internal pressure resulting from the forces that hold the crystal lattice together. The total pressure thus acting on a solid body may be represented by $p + a$, the sum of the external and internal pressures. Moreover, if we plot $p$ as a function of $T$ and produce the curve below the triple point, we see that it cuts the $p$-axis at $-a$. $a$ may thus be interpreted as the negative pressure at which the substance would melt at zero temperature. We thus see that the dependence of melting pressure on temperature is in the main given by the constant $c$. As Simon showed, this constant varies comparatively little from one substance to another and is roughly equal to 2. Writing $p/a = p'$, the formula may thus be expressed in the reduced form

$$p' = T'^c - 1, \qquad c \sim 2. \qquad \ldots\ldots(3)$$

Now it is evident that if $a$ really signifies the internal pressure, $a$ will be small for such substances as melt at low temperatures. For a given value of $p$ the reduced pressure $p'$ will therefore be the greater the lower the triple point of the substance. Similarly, for a given temperature $T$ the reduced temperature $T'$ will be the greater the lower the triple point. Thus if our physical interpretation of the formula is correct, we may conclude that in order to follow up the melting curve as far as possible it will be advantageous to study substances with low triple points. For with these substances we may, by employing the pressures at our disposal, attain the greatest reduced pressures and simultaneously the greatest reduced temperatures.

It was these arguments that induced Simon and his collaborators to investigate the melting curves of condensed gases at low temperatures.

### I. VI. 2.  *The Melting Curves of Condensed Gases*

It was a daring step of Simon to choose helium as his first object of research. All the characteristic data of helium, critical temperature and pressure, boiling point, etc., are indeed lower than in the case of any other substance. On the other hand its triple point is non-existent. All attempts of Kammerlingh Onnes to solidify helium by lowering the temperature along the vapour-pressure curve failed, and it was not until after Onnes' death that solid helium was discovered. Keesom found that the melting curve of helium, far from meeting the vapour-pressure curve at a triple point, remains at a pressure of about 25 atm. down to the lowest temperatures attainable. From the shape of the curve Keesom concluded that the melting pressure does not fall to zero at zero temperature but intersects the pressure axis horizontally at about 20 atm. This behaviour, which is quite different from that of any other substance, is also completely at variance with Simon's formula. In spite of this fact, which makes it impossible to calculate the melting pressure at higher temperatures, Simon began his experiments by trying to solidify helium at temperatures attainable with solid and liquid hydrogen (10–20° K.), employing pressures between one and two thousand atmospheres.

These experiments were successful, and as soon as the first melting points were found there was no great difficulty in following up the curve. In the course of subsequent work Simon and his collaborators were able to solidify helium at temperatures up to 40° K. with pressures up to 6000 atm.

The simplest method to determine solidification in a liquid is to drive it into a U-shaped tube fitted with a manometer at each side of the bend, the latter being cooled to a definite temperature (fig. 2). If the pressure be then gradually increased, the solidification pressure at the temperature in question is made evident by the fact that a further increase of pressure moves only the manometer nearest the pump. The solidified substance at the bend ceases to transmit the pressure to the second manometer. This method was employed by Simon in most of his experiments. In order to avoid dangerous explosions at high pressures, due to bursting tubes, it was considered advisable to use very small volumes of helium. The gas (fig. 3) was pressed by a mercury pump $R$ into a very narrow steel capillary containing the U-shaped tube with its bend at $E$. As the volume of an ordinary Bourdon manometer would have been too large and would moreover have necessitated joints, which are difficult to make gastight at high pressures, the manometers were re-

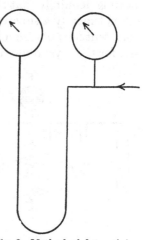

Fig. 2. Method of determining solidification at high pressure.

placed by spirals $M$ wound from the same capillary tubing and fitted with small mirrors $I$. The spirals are deformed by the pressure in the capillary and the position of the mirrors can be read off with a scale and telescope. The sensitivity of these spiral manometers can be raised at will by increasing the number of coils. In order to increase the pressure range beyond the capacity of the pump, a reservoir spiral $L$ was inserted, which could be cooled with liquid air. Mercury was driven up the pump as far as the loop $F$, which was then cooled so that the mercury froze, thus cutting off the pump from the apparatus. Thereupon the liquid air was removed from $L$. As a result of the expansion of the gas in $L$ the pressure could be raised by some 50 per cent., if the dimensions of the various spirals were suitably chosen.

In the first experiments another method was employed, in which the melting temperature at a given pressure was measured by noting at what point a break occurred in the heating curve of the solid. The gas was pressed into a small steel cylinder and there solidified. The temperature was then allowed to rise gradually and noted as a function of time. When the melting point was reached the temperature remained constant during fusion. This method has the advantage of showing definitely that the solid phase was indeed crystallised and not merely a supercooled viscous liquid. The appearance of a latent heat of fusion at a definite temperature is characteristic for the crystalline state.

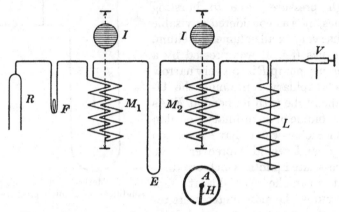

Fig. 3. Apparatus for determining melting curves of condensed gases at high pressures.

Apart from the experiments on helium Simon studied the melting curves of a number of other condensed gases. His results may be summarised as follows:

(1) The melting curves of He, $H_2$, Ne, $N_2$ and A may all be expressed by the formula $\log(p+a) = c \log T + b$, with the exception of He below $3°$ K.

(2) With pressures up to 6000 atm. the melting curve of He may be followed up to $42°$ K. This temperature is twenty times as high as the triple point would be if the curve remained regular at low temperatures, ten times as high as the normal boiling point and eight times as high as the critical point of helium.

(3) The temperatures attainable on the melting curves of the other substances are not so high in relation to the triple points and other characteristic temperatures.

(4) No signs were found of critical phenomena between the solid and the fluid phases.

In fig. 4 the melting curves of helium, hydrogen, neon, nitrogen and argon are shown as taken from Simon's measurements. The dotted parts are extrapolated. On each curve the critical point is marked by a small square for comparison.

Though we have obviously no right to extrapolate these results, the melting curve of helium clearly shows that the solid phase can exist at temperatures far above those at which

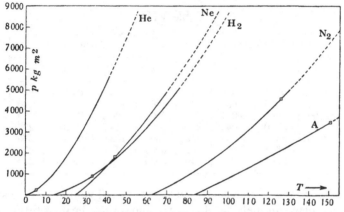

Fig. 4. Melting curves of He, $H_2$, Ne, $N_2$ and A.

the difference between gas and liquid ceases. The liquid phase as distinct from the gas occupies only a minute portion of phase space as compared with the solid. As no difference is apparent in the behaviour of solid helium above 3° K. and that of all the other substances investigated, helium at 40° K. and 5500 atm. may, in accordance with the principle of corresponding states, be considered as representing, at any rate qualitatively, the behaviour of nitrogen, argon, etc. at progressively higher temperatures and pressures. We may thus infer with considerable justification that the shape of the melting curve of nitrogen at temperatures some twenty times its melting point (1260° K.) will not be very different from that of helium at 42° K. Simon himself drew attention to the fact that if we assume the melting curves of metals and minerals not to differ too much from those of other substances, as for instance experiments on mercury enable us to infer, important conclusions can be drawn as to the state of bodies in the interior of the earth and planets.

The question of the critical point between the solid and fluid phases was left open as a result of these experiments. All that could be shown was that if there is such a point, it must be at exceedingly high pressures. Some further information was obtained by recent results of Bridgman on nitrogen and argon. Bridgman determined the melting pressures of these substances

Table VIII

| Nitrogen | | | | Argon | | | |
|---|---|---|---|---|---|---|---|
| $p$ | $T$ | $\Delta V$ | $\lambda$ | $p$ | $T$ | $\Delta V$ | $\lambda$ |
| 1 | 63·14 | (0·072) | (218) | 1 | 83·9 | 0·0795 | 280 |
| 1000 | 82·3 | 0·058 | 271 | 1000 | 106·3 | 0·0555 | 280 |
| 2000 | 98·6 | 0·047 | 302 | 2000 | 126·4 | 0·0425 | 279 |
| 3000 | 113·0 | 0·040 | 334 | 3000 | 144·9 | 0·0340 | 277 |
| 4000 | 125·8 | 0·033 | 335 | 4000 | 162·0 | 0·0280 | 275 |
| 5000 | 137·8 | 0·029 | 342 | 5000 | 178·0 | 0·0240 | 276 |
| 6000 | 149·2 | 0·026 | 346 | 6000 | 193·1 | 0·0210 | 277 |

up to 6000 atm., together with the change of volume, $\Delta V$, that takes place on melting. From these data the heat of fusion $\lambda$ could be computed with the help of the Clausius-Clapeyron equation. Table VIII shows Bridgman's results.

If there were a critical point between the two phases, $\Delta V$ and $\lambda$ would have to vanish together there. Table VIII shows that, at any rate at these pressures and temperatures, there is no sign of this occurring. $\Delta V$, it is true, diminishes but does not appear to be tending to zero at any finite temperature, whereas $\lambda$ in the case of nitrogen increases and in the case of argon is practically constant. Bridgman concludes that there is no critical point and that the melting curve continues indefinitely towards higher temperatures and pressures.

According to Landau, there can be no critical phenomena between the crystallised and liquid phases, since the transition is necessarily connected with a change in symmetry.

The results of Bridgman and Simon, though qualitatively

in agreement, show appreciable quantitative deviation. As Bridgman has not yet published details of his apparatus, it is difficult to gauge the accuracy of his results. Simon's measurements were not particularly accurate and were intended rather to obtain a survey over the entire field. Pressures were read on a Bourdon gauge that could not be calibrated as no possibility existed at the time in Europe. The pressures deduced from the spiral manometers were extrapolated from those calibrated with the Bourdon gauge by assuming the validity of Hooke's law. Simon's equation satisfies Bridgman's figures, but the constants $a$, $b$ and $c$ differ.

At low pressures accurate measurements have been made by Keesom and Lisman on hydrogen up to 600 atm. and on neon and nitrogen to 200 atm., using the method of the plugged capillary. The results of these experiments are in accord with neither Bridgman nor Simon. In the region where the points overlap, considerable deviations occur. According to these authors, Simon's formula may be used merely as a fair interpolation, but in actual fact it does not represent the facts accurately enough. Moreover, the values deduced for the constants from these low-pressure measurements differ very considerably from those given by the others. A summary of all the results is given in the following table:

Table IX. *Melting data*

$$\log(p+a) = c \log T + b$$

| Name | $a$ | $b$ | $c$ | Observer |
|---|---|---|---|---|
| Helium | 17·0 | 1·236 | 1·5544 | Simon, Ruhemann, Edwards |
| Hydrogen | 220·0 | 0·191 | 1·879 | Simon, Ruhemann, Edwards |
| | 245·3 | 0·28771 | 1·83435 | Keesom, Lisman |
| Neon | 1200·0 | 0·875 | 1·579 | Simon, Ruhemann, Edwards |
| | 728·8 | −0·16875 | 2·18038 | Keesom, Lisman |
| Nitrogen | 1250·0 | −0·870 | 2·203 | Simon, Ruhemann, Edwards |
| | 933·4 | −2·527384 | 3·05365 | Keesom, Lisman |
| | 1600·0 | −0·0634 | 1·815 | Bridgman |
| Argon | 3000·0 | 1·003 | 1·288 | Simon, Ruhemann, Edwards |
| | 2500·0 | 0·5546 | 1·478 | Bridgman |

The constants ascribed to Bridgman were determined from his published results.

RLTP ⟨ 97 ⟩ 7

### I. VI. 3. *Melting Diagrams of Binary Solids*

The melting diagrams of binary mixtures correspond to the boiling curves and dew curves of gaseous-liquid systems. Just as complications occur in the latter when the components are

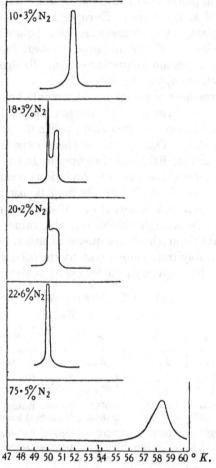

Fig. 5. Specific heat of $O_2$-$N_2$ mixtures in the solid-liquid region.

not completely miscible in the liquid state, the equilibrium curves of melting and solidification differ according to the miscibility relations in the solid.

The melting diagrams of metals have long been a source of interest to metallurgists, but comparatively little is known of non-metallic substances. More especially the melting diagrams of solidified gases were until recently completely unexplored. Even the comparatively simple case of solid air, which everyone working at low temperatures has frequently observed in a Dewar flask after pumping off the vapour, had never been analysed till 1934. It was not known whether these crystals consisted of pure nitrogen or of a eutectic mixture or whether they were solid solutions. Very little systematic work has been done to ascertain the laws governing the formation of mixed crystals. From a study of the data on metals one is tempted to lay down a number of empirical rules, but none of them will stand much criticism. All we can say at present is that whether solid solutions are formed or not depends on the physical properties of the components, their crystal symmetries and lattice constants. But we can hardly make even a qualitative guess at the shape of the melting curve of a given pair of substances.

The case of liquefied gases is of technical as well as scientific interest. For all processes carried out at low temperatures, such as rectification of coke gas, etc., industry is dependent on liquid cooling agents. In many cases it is advantageous to employ the lowest available temperatures. The lower limit of temperature attainable with a given liquid is given by the freezing point; for no technical plant can as yet work with a solid cooling agent. The tubes immediately become blocked and the apparatus ceases to work.

Now it is a well-known fact that in cases when eutectic solutions are formed the freezing point of a mixture is invariably lower than that of both components. It is therefore possible that a systematic study of the melting diagrams of binary systems of solidified gases may disclose cooling agents that will not freeze down to very low temperatures.

With this end in view Ruhemann and Lichter at Kharkov studied several such mixtures with a calorimetric method. This has considerable advantages as compared with the usual method of thermal analysis employed at high temperatures, in which the specimen is allowed to cool slowly and the temperature-time curve plotted. In the latter the formation of two phases is registered by a kink in the curve, which is often inadequately defined, and little or no quantitative results may be obtained concerning the heats of fusion. By measuring the specific heats as a function of temperature a much clearer picture is obtained. This may be illustrated by fig. 5, which shows the specific-heat-temperature relation for various mixtures of oxygen and nitrogen. The diagram deduced from the results is given in fig. 6. We see that in two of the

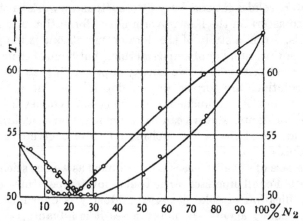

Fig. 6. Melting diagram of oxygen-nitrogen mixtures.

curves in fig. 5 the specific heat shows two maxima of rather different shapes. One forms a sharp and narrow peak almost like a heat of transition and has the same position though a different height for each composition; the other is flatter and its position on the temperature axis depends on the composition of the mixture. The first peak is clearly due to the melting of a eutectic mixture, which in the ideal case should occur sharply at one and the same temperature for every mixture outside the limits of solubility; in the second hump the rest of the mixture melts along a range of temperature, the extent and position of which is dependent on the composition. If we bear this in mind it is easy to deduce fig. 6 from fig. 5. We see that oxygen and nitrogen are extensively soluble in one another. Only a comparatively narrow range of concentrations between 16 and 30 per cent. $N_2$ is taken up by the solubility gap. Evidently up to 16 per cent. of nitrogen may be dissolved in the oxygen lattice, whereas 70 per cent, of oxygen enters the lattice of nitrogen. The eutectic temperature is 50·1° K. and the eutectic concentration 22 per cent. $N_2$. That is to say, a mixture of this composition may be cooled to 50·1° K. without the formation of a solid phase. Unfortunately it appears difficult to obtain this temperature

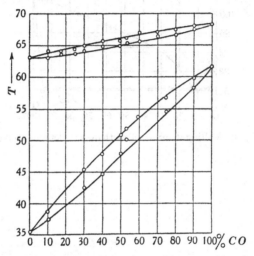

Fig. 7. Melting diagram of nitrogen-carbon monoxide mixtures.

by pumping off the vapour; for the vapour pressure at the quadruple eutectic point is 1·2 mm. of mercury. It needs a strong pump to maintain this low pressure over a liquid surface.

Apart from oxygen-nitrogen mixtures the authors studied mixtures of nitrogen and carbon monoxide and of methane and ethylene. The diagrams are shown in figs. 7 and 8. The methane-ethylene system has a eutectic point at 84·5° K. and 12·2 per cent. $C_2H_4$. Only about 2·5 per cent. of ethylene is soluble in solid methane at the melting point. The solubility of methane in ethylene is also small but could not be determined accurately.

The system $N_2$-CO shows complete solubility in the solid

phase, the non-homogeneous lens being very narrow, not more than 1° at the widest point. This might have been expected, as the two molecules are physically very similar and the crystal lattices are almost identical (see Part II, chap. I). In the figure the transition lines between the α- and β-phases are also shown.

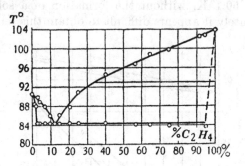

Fig. 8. Melting diagram of methane-ethylene mixtures.

# Part II

## THE SOLID STATE

---

### CHAPTER I

### THE CRYSTAL LATTICE

#### II. i. 1.  *X-ray Methods at Low Temperatures*

Until about ten years ago the problem that the solid state presented to physicists consisted in the determination and correlation of the material constants of a given solid. This problem found its solution in the highly developed theories of Debye, Born and Kármán. It received its main impulse from the knowledge that a crystal is a periodic lattice and that the lattice points are capable of taking up vibrational energy in quanta. It introduced these conceptions step by step into the continuum theory of solids and thus considerably refined the classical model. It is, however, essentially a classical theory and, in spite of its perfection, subject to all the limitations of the classical outlook.

Some of these limitations were felt early. Application to crystals of lower symmetry than cubic proved to be extremely difficult, the difficulties rising to impossibility with increasing complexity of the chemical and crystal structure. Cubic crystals of monatomic substances or of compounds containing few chemically different atoms or groups were therefore considered simple, and the relative abundance of such cases was regarded as a particularly fortunate circumstance.

Curiously enough, for a time it was felt that to low-temperature research should fall the task of revealing more simple cases; it was even hoped that the reduction of thermal agitation would, generally speaking, leave matter in a simplified state which would lend itself more readily to theoretical interpretation.

The curiosity of research workers, who in spite of these unattractive prospects probed into the experimentally difficult realm of low temperatures, was richly rewarded. The reduction of thermal motion, far from making matters simpler, revealed instead a whole world of interesting phenomena which, although of course in themselves not a feature merely of low temperatures, are at normal and higher temperatures less prominent and hence not so accurately discernible. The immediate consequence of these discoveries was the development of precision methods particularly suited to low-temperature research. This task has been attacked with much energy and success, so that at the present moment it is possible to determine physical quantities with the same thoroughness and accuracy as under normal conditions, and in some fields with an accuracy exceeding any methods developed in the regions of normal and higher temperatures.

The determination of crystal forms stable at low temperatures is a very conspicuous example of this development. Only a few years ago, when X-ray methods had already been highly developed and had been applied to the location of the last atom in the most complicated structures to an extent which is best demonstrated by the size of the "Strukturbericht", the low-temperature physicist was still content with the statement that a crystal belonged to the cubic, hexagonal, tetragonal or else a more complicated system and with the determination of lattice constants to an accuracy of several per cent. The reason lies in the experimental difficulties of producing and handling crystals of size and form suitable for accurate investigation at low temperatures.

In many cases it is easiest to obtain a crystal powder, which can be examined by the Debye-Scherrer method only; in other cases even this simple method is inapplicable, because the powder is not sufficiently finely grained. The investigation of single crystals has only recently become possible through the application of new methods. Optical and electrical methods have also been used, though not as widely as X-rays; indeed, the earliest attempts at obtaining information about the symmetry of crystals formed by gaseous substances on cooling were made by Wahl in 1914 by means of a polarisation microscope. He examined a large number of substances in a small quartz vessel with plane parallel walls immersed in a Dewar vessel with liquid hydrogen. His results have, however, in many instances not been confirmed by later measurements.

The preference for the Debye-Scherrer method is quite natural; the adaptation to low-temperature research requires no essential changes in the principle. However, the necessity of insulating the specimen thermally from the surrounding camera offers many technical difficulties.

The most primitive construction, which is useful at temperatures above 100° abs., consists in a wrapper of insulating material packed around the camera, through which is passed a current of cold gas, previously freed from easily condensed impurities such as water, $CO_2$, etc. The only advantage of this method is that no vacuum pump is needed; since in most laboratories a high vacuum is not considered a real difficulty, this method cannot now be considered a very satisfactory solution.

The obvious advantages of the evacuated camera have led to its predominant use in the whole temperature region from 0° C. to 2° K. In principle it consists of a metal box with two openings for transmitting the X-ray, on the top of which is placed the cooling device to which the crystal is fixed. The box also contains the film carrier.

The details of construction depend on the nature of the crystals and the temperature region in which they are to be examined, as well as on the choice of the cooling device. The cameras in use are therefore extraordinarily varied and it would be impossible to describe all their different advantages. We shall confine our description to a few essentials.

If the substance under ordinary circumstances is gaseous or liquid it is usually deposited on the outside of a thin-walled capillary or on a wire. These must be kept well in the centre of the camera; as on cooling the different parts of the apparatus contract differently, it is necessary to introduce adjusting screws. These can be screwed into a warm part of the apparatus, but must then have tips of fibre, to avoid heating the specimen, as illustrated in fig. 1 ($M_1$, $M_2$, $M_3$). If glass is used for the capillary, it can simply be fused to a double-walled glass vessel, containing the cooling liquid, which in turn is connected to the metal camera by means of a conical ground joint or "picein". The gas is usually admitted directly into the camera; if, however, at the desired temperature the substance has a considerable vapour pressure or if the gas is likely to attack the metal parts of the camera, the space around the capillary must be

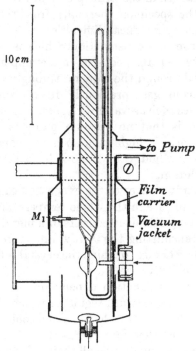

Fig. 1. X-ray camera for condensed gases with high vapour pressures.

isolated from the vacuum of the camera. This can be effected as in fig. 1 by means of a very thin-walled glass bulb.

Metal capillaries are used when it is feared that gas bubbles originating at the lower end of the capillary prevent the specimen from obtaining a sufficiently low and well-defined temperature; to effect intense and even cooling it is advisable to let the cold gas or liquid pass continually through the capillary (fig. 2).

For many purposes it is sufficient to use a metal wire instead of a capillary, since the thermal conductivity of pure metals, especially in these temperature regions, is considerable. The use of metal is to be recommended in all cases where the cooling device is itself of metal. Although more complicated in construction, the latter is particularly advantageous for work at very low temperatures, especially in the helium region. Here it is indispensable that the specimen be very carefully screened off from temperature radiation, and therefore the metal frame carrying the photographic film must have the

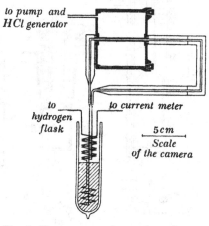

to pump and
HCl generator

to
hydrogen
flask

to current meter

5 cm
Scale
of the camera

Fig. 2. X-ray camera for condensed gases.

same temperature as the specimen. Fortunately the sensitivity of the photographic layer to X-rays is only slightly reduced by cooling. Fig. 3 shows an all-metal camera constructed by Keesom.

If the specimen is obtained in crystal form at room temperature it is sometimes difficult to find a suitable way of bringing it into good thermal contact with the cooling agent. Metal films can be sputtered on a thin layer of transparent material, e.g. celluloid or acetyl cellulose mounted on a metal frame, which is soldered to the cooling apparatus. The thermal conductivity of the metal is usually sufficient even in a thin layer to guarantee uniform distribution of temperature. A crystal powder of bad thermal conductivity is best filled in a thin-walled glass ampulla together with hydrogen or helium gas and the ampulla sealed off. Wood's alloy, if prepared from pure ingredients and *in vacuo*, will make a reliable thermal joint to the metal of the cooling device, and it does not break the ampulla on cooling.

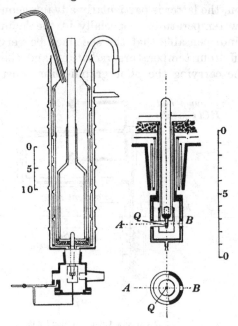

Fig. 3. All-metal X-ray camera.

Acetyl cellulose is a very suitable material for the X-ray windows of the camera. It neither absorbs nor scatters X-rays to an appreciable degree and at a thickness of 12 or even $6\mu$ is perfectly vacuum-tight. It withstands atmospheric pressure on a considerable area. It can therefore be used to eliminate the one great drawback of the vacuum camera, i.e. that the photographic film is placed inside the evacuated box. Fig. 4 shows a camera with an X-ray window $G$ surrounding the specimen over an angle of more than 270°. Only the central tube $R$ with this window is evacuated and the film carrier $B$ is arranged outside the vacuum, so that long series of measurements can be carried out on the same specimen without destroying the vacuum or heating up in between. In combination with a continuous temperature regulator and a powerful X-ray tube, a camera of this type enables us to study in closest detail transitions in crystals at any given thermal treatment.

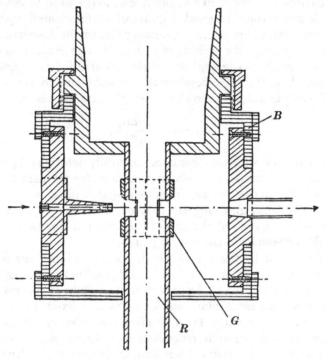

Fig. 4. X-ray camera with film outside vacuum.

The exclusive use of the Debye-Scherrer method is of course a serious limitation. The Seemann-Bohlin method has occasionally been used, but it is not very convenient at low temperatures; moreover, it gives us no more information than the former. A decisive advance was not made until single-crystal methods were adopted. This is an urgent problem with respect to substances which at room temperature are not obtainable in crystal form and the structure of which cannot be identified from powder diagrams. The chief difficulty involved is the production of the single crystals, and it has as yet been solved only for modifications stable directly below the melting point. And even this solution did not prove quite satisfactory: the crystals formed by the liquid are rarely oriented in a rational axis. Since it would be too difficult to orient the crystal grown in an accidental orientation, which is naturally enclosed in a vacuum, methods had to be devised to decipher the rotational diagram of an unoriented crystal. For crystals of not too low symmetry this can be done without great difficulty: the position of a reflection on the film is determined by the Bragg angle $\theta$ and the "Schichtlinien" angle $\mu$. From these the angle between the crystal plane, to which the reflection is due, and the axis of rotation are calculated by

$$\cos \xi = \frac{\cos \theta \tan \mu}{\tan 2\theta}.$$

One can now compare these experimentally determined $\xi$ with those calculated for a crystal system to which we more or less reasonably expect the specimen to belong. Unless we are very unfortunate in guessing, the method leads fairly quickly to the determination of the axis of rotation and the correlation of the reflections to the crystal planes.

To eliminate the guesswork, Keesom combined the rotational motion of the crystal with a longitudinal motion of the film in the direction of the axis of rotation. Two exposures are taken on the same film, one with the film fixed, the other with the film moving. The first gives the ordinary rotational diagram with the four reflections from every crystal plane arranged at the corners of a rectangle symmetrically situated to the equator and perpendicular to it through the point of intersection of the primary X-ray beam with the film. The diagram with moving film again contains the same four reflections from each crystal plane, but the rectangle is drawn out into a rhombus in the direction of the motion of the film. Knowing the distance through which the film has been moved and the angle through which the crystal has been rotated during the exposure from the relative positions of the corresponding rectangle and rhombi, the angles between any two crystal planes can be determined directly.

Figs. 5 and 6 represent two cameras devised for the

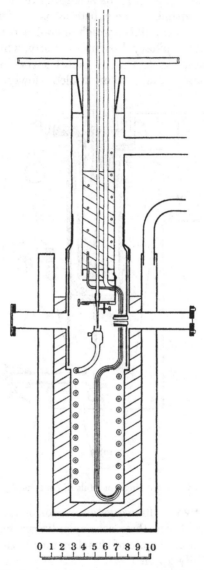

Fig. 5. X-ray camera for single crystals of
condensed gases (Ruhemann).

investigation of single crystals according to the principles we have roughly outlined. The substance is condensed in liquid form in the glass capillary, which is cooled from below. The lower end of the capillary dips into mercury, which on freezing closes it vacuum-tight. On this metal surface single crystals are grown fairly easily when cooled slowly. However, it

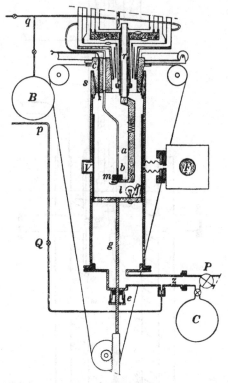

Fig. 6. X-ray camera for single crystals of condensed gases (Keesom).

happens that two or three crystals are formed; unless they are of very different size their corresponding reflections are not easily told apart. In spite of this difficulty both cameras have been successfully applied.

Another recent development is the refinement of precision in X-ray measurements at low temperatures. The main interest here centres in the determination of expansion coefficients.

The elaborate dilatometer apparatus developed for use at high temperatures is extremely unpractical at low temperatures; thus in a great many cases X-ray methods have proved the best way out of the difficulty. Although the most refined methods of modern X-ray technique, yielding an accuracy equal to that of the best dilatometer, have as yet not been applied, the simple Debye-Scherrer method has been used with good success. It is capable of an accuracy of about 0·5 X-units

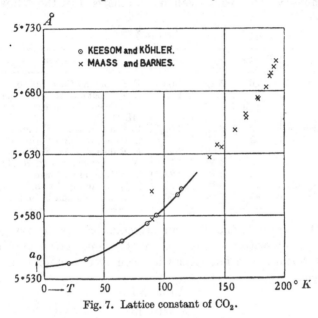

Fig. 7. Lattice constant of $CO_2$.

or approximately one-hundredth of a per cent. in the determination of the lattice constant, which for many low-temperature problems is already quite useful. This accuracy is obtained by measuring large deflection angles $\theta$ and extrapolating the values of the lattice constant, obtained from various $\theta$, to $\theta = 90°$ according to Kettmann's method. In this manner two useful standards for low-temperature measurements of lattice constants have been obtained by Keesom and Köhler, silver and carbon dioxide. The lattice constant of silver at 83·2° K. equals $4·068 \pm 0·0004$ Å.; that of carbon dioxide is plotted as a function of temperature in fig. 7.

## II. 1. 2.  *Crystal Structures Stable at Low Temperatures*

The only truly simple cases which the extension of structural analysis to the region of low temperatures has divulged are the crystals of the *rare gases*. With the exception of helium and emanation, which owing to experimental difficulties have not yet been investigated, they have been found to possess the cubic face-centred lattice which becomes atoms of spherical symmetry. The results can be summarised in the following table:

### Table X

| Name | $T°$ abs. | $a_w$ Å. | $\rho$ | Observer |
|------|-----------|----------|--------|----------|
| Argon | 40 | $5.4_2$ | 1.65 | F. Simon and Cl. v. Simson |
|  | 20 | $5.4_0$ | 1.68 | I. de Smedt and W. H. Keesom |
| Neon | ~4, 5 | 4.52 | . | I. de Smedt, W. H. Keesom and H. H. Mooy |
| Krypton | 82 | $5.68_4$ | $3.00_4$ | B. Ruhemann and F. Simon |
|  | 89 | $5.69_4$ | $2.99_4$ | B. Ruhemann and F. Simon |
|  | 92 | $5.69_7$ | $2.98_6$ | B. Ruhemann and F. Simon |
|  | 20 | 5.59 | 3.13 | W. H. Keesom and H. H. Mooy |
|  | 78 | 5.78 | 2.83 | G. Natta and G. Nasini |
| Xenon | 100 | 6.18 | 3.70 | G. Natta and G. Nasini |
|  | 88 | 6.24 | 3.56 | B. Ruhemann and F. Simon |

The results of the different investigators are in very good agreement, with the exception of the values given by Natta and Nasini; it is not clear why their results differ from the others.

### Table XI

$$T = 100° \text{ K.}$$

| Metal | Lithium | Sodium | Potassium | Rubidium | Caesium |
|-------|---------|--------|-----------|----------|---------|
| $a_w$ | 3.46 | 4.24 | 5.25 | 5.62 | 6.05 |

The final determination of the crystal structures of the *alkali metals* also necessitated the application of low temperatures, since these metals reflect X-rays at room temperature very badly. F. Simon and E. Vohsen examined thin foils of all the alkalis at 100° abs. All foils gave fine, intensive Debye-Scherrer lines, with the exception of lithium, which had to be hammered in liquid air to assume a finely grained texture.

All the diagrams corresponded to cubic body-centred lattices, the constants of which were determined with an accuracy of 0·5 per cent. as given in Table XI. No changes of crystal structure could be observed.

*Mercury* has been the object of very extensive study. The earliest experiments on its crystal structure, as reported in the "Strukturbericht", led to contradictory results, the evidence in favour of the primitive rhombohedric cell, first proposed by McKeehan and Cioffi, being confirmed by later experiments as against the hexagonal lattice suggested by Asen and Aminoff. It was examined in the form of minute droplets, condensed in a capillary and frozen by means of liquid air or solid $CO_2$, or as a layer evaporated on the cooled surface. The final evidence, however, was gained from rotational diagrams of single crystals obtained in the camera described above (fig. 5). It was found, after indexing the diagram, that the crystal had grown in an irrational axis, almost coinciding with (3 3 $\bar{1}$). All the reflections that were to be expected from the structure proposed by McKeehan and Cioffi were indeed found on the diagram and no more. It must thus now be considered as fairly certain that mercury crystallises in the rhombohedric class with only one atom in the elementary cell; from the coincidences of several reflections it must be inferred that the angle $\alpha$ of this cell is determined by

$$\cos^{-1}\alpha = 1/3 \qquad (\alpha = 72° 32').$$

The lattice constant is 3·0 Å., but has not been determined very accurately.

Mercury is the last of those elements, the structures of which are determinable only with the help of low temperatures, which crystallises in a comparatively simple monatomic lattice. Chlorine and bromine have not as yet been examined, but there is little hope that they are of a simple type.* The remaining elements are the diatomic gases $H_2$, $O_2$, $N_2$ (with them we must also class CO, as being almost identical with $N_2$). With the possible exception of hydrogen, all of them exhibit more than one crystal form and all have been studied though not with equal success. They offer considerable technical and experi-

[1] See addenda to bibliography for the structure of chlorine.

mental difficulties. Perhaps the most extraordinary perform-
ance was the determination of the crystal structure of hydrogen.
The scattering power of hydrogen is, of course, extremely small;
moreover, the crystals do not give very good powder diagrams.
The sample has to be cooled down to the temperature of
liquid helium. Keesom and his collaborators obtained their
best diagram, showing eight lines originating from the $H_2$
lattice at $1.65°$ K., with an exposure of 5 hours (Fe radiation,
26 kV., 10 mA.) from a sample 3 mm. thick. The result, in
agreement with those derived from two other exposures,
indicated that hydrogen crystallises in a hexagonal close
packing with two molecules in the elementary cell, the
dimensions of which are

$$a = 3.75 \text{Å.},$$

$$c/a = 1.633,$$

$$\rho = 0.088 \text{ at } T = 1.65° \text{ K.}$$

These data are for pure parahydrogen, which down to that
temperature suffers no transformation.

The result seemed surprising, since Wahl had failed to discover double refraction in solid hydrogen under the polarisation microscope. Later it was found that the hexagonal close
packing is rather preferred by diatomic molecules. CO as well
as $N_2$ have modifications of this type. The so-called $\beta$-modifica-
tion of $N_2$ was identified by means of single-crystal rotational
diagrams, as good powder diagrams were difficult to obtain;
the $\beta$-crystals are stable near the melting point, where the
vapour pressure and the speed of recrystallisation are still high.
The powder diagrams obtainable corroborate the structure for
both $\beta$-$N_2$ and $\beta$-CO. This structure has a striking peculiarity.
The co-ordinates of the atoms cannot be derived from the line
intensities of the diagram; only the centre of the molecules can
be located. The molecule behaves as though of complete
spherical symmetry. In the case of hydrogen this is most
probably due to a rotational motion of the molecule in the
crystal lattice; calculations based on the specific heat curves
of $N_2$ and CO, carried out by Giauque, show that this conception can hardly be applied to the latter two substances.

The complete analogy between the lattices of CO and $N_2$ is best seen from a comparison of the respective data:

Table XII

|      | $a$   | $c/a$  | $\rho$ | $T$      | Observer  |
|------|-------|--------|--------|----------|-----------|
| $N_2$ | 4·039 | 1·651  | 0·982  | 45° K.   | Vegard    |
| CO   | 4·11  | 1·651  | 0·929  | 65° K.   | Vegard    |
| $N_2$ | 4·034 | 1·633  | 0·995  | 39° K.   | Ruhemann  |

When cooled to lower temperatures the two substances behave similarly. They both undergo an allotropic transition, $N_2$ at 35·5° K., CO at 61·5° K. Again the modifications stable below these temperatures are identical; according to Vegard they are cubic, almost a close-packed structure, but with a characteristic complication: the centre of the molecule seems not to coincide with the corners of the cube, but appears to be slightly displaced in the direction of the trigonal axis, so that the space group is $T^4$. That the CO molecule should thus exhibit a slight polarity is not astonishing; apparently the $N_2$ molecule does so too. The evidence depends, however, on two rather weak lines, which could not be found on a diagram taken by Ruhemann on $N_2$, although from their relative intensity they ought to have been observed.

An entirely different structure has been proposed by Keesom and his collaborators for $N_2$; they find agreement with the diagram of Vegard, but they show that a tetragonal cell containing four atoms can explain the diagram with the same accuracy as the cubic cell with eight molecules proposed by Vegard. As observation under the polarisation microscope had revealed double refraction of $N_2$ at the temperature of liquid hydrogen, the authors prefer the tetragonal description. We summarise these various results in Table XIII.

Table XIII

|      | $a$   | $c/a$      | $n$ | $d$   | $(u+v)/2$ | $\rho$  | $T$     | Observer                        |
|------|-------|------------|-----|-------|-----------|---------|---------|---------------------------------|
| CO   | 5·63  | 1·0        | 4   | 1·065 | 0·03      | 1·0288  | 20° K.  | Vegard                          |
| $N_2$ | 5·66  | 1·0        | 4   | 1·065 | 0·015     | 1·0265  | 20° K.  | Vegard                          |
| $N_2$ | 5·67  | 1·0        | 4   | ·     | 0·0       | ·       | 20° K.  | Ruhemann                        |
| $N_2$ | 5·656 | $\sqrt{2}$ | 2   | ·     | ·         | 1·024   | 20° K.  | De Smedt, Keesom and Mooy       |

Oxygen had until recently resisted all attempts to resolve its crystal structure. It is by far the most complicated of this group, existing in three different modifications, $\alpha$, $\beta$, and $\gamma$. The transition temperature from $\alpha$ to $\beta$ is very low, 23·5° K.; that from $\beta$ to $\gamma$, 43·8° K. The $\alpha$-modification has been investigated several times; McLennan suggested an orthorhombic structure, which however could not be corroborated by subsequent investigators. Ruhemann arrived at the consequence that the symmetry certainly cannot be higher than rhombic, whereas Mooy suggests a hexagonal cell with twelve atoms as a possibility. The diagrams are very rich in lines. As in most investigations of condensed gases, the sample has to be of considerable diameter (of the order of 1 to 3 mm.) in order to shorten the exposure, and owing to the transparency of these light atoms, the inner lines of the diagrams are split into two components. This makes the indexing still more difficult. Unfortunately there are as yet no methods known for producing single crystals of modifications stable below a transition point; even if a $\gamma$ single crystal is obtained, it breaks up into very many small crystals immediately after cooling through the first transition temperature. Thus the $\beta$-modification is also accessible only by way of Debye diagrams. Such have been obtained by Ruhemann, but nothing could be derived from them except that they differ from the diagrams of $\alpha$-$O_2$ only very slightly. Subsequently Vegard succeeded in describing his and Ruhemann's Debye-Scherrer diagrams with the help of a rhombohedric cell containing 6 molecules and was able to determine the structure of the $\gamma$-modification slightly below the melting point. The structure was found to be cubic with the space group $T_h^6$. The cell contains 8 molecules, which appear to be of spherical symmetry as in the case of $\beta$-$N_2$ and $\beta$-CO. According to Vegard, the molecules are arranged in pairs. The distance between 2 molecules of a pair is 4·48 Å., whereas the shortest distance between two pairs is 4·68 Å.

Although some unsolved problems remain, we see that low-temperature research has contributed appreciably to our knowledge of the crystal structures of the elements. Ample information has further been obtained for a number of compounds of simple composition that are gaseous at room temperature, and many crystals have revealed new characteristics at low temperatures. Progress in this field has been somewhat slower, unexpected complications being encountered which have not yet found a really satisfactory explanation, but which revealed extremely interesting new aspects of matter.

The first group with which we will deal here are the hydrogen halides: HCl, HBr, HI. HCl exists in two modifications, one of very low symmetry which could not be determined, stable at low temperatures, and one stable between 98° K. and the melting point. This is a cubic close packing

$$a = 5 \cdot 5 \pm 0 \cdot 05 \, \text{Å.},$$
$$\rho_{107°} = 1 \cdot 469,$$
$$\rho_{81°} = 1 \cdot 50_7.$$

Of course the data obtained from X-ray diagrams refer only to the centre of the molecule; as the scattering power of the H-atom is quite negligible in comparison with the halide atom, the high symmetry leading to the close-packed structure is evidently not due to a corresponding symmetry of the molecule itself, as the low-temperature modification shows. Simon, who investigated the structure, pointed out that it might be due to a rotational motion of the HCl molecule, owing to which it obtains spherical symmetry. He corroborated his argument by referring to the higher specific heat of the cubic modification at the transition temperature (fig. 8). Later experiments of Cone, Dennison and Kemp, and Smythe and Hitchcock showed that the dielectric constant of HCl, which changes only very slightly at the melting point, drops sharply to a very low value at the transition temperature (fig. 9); this result can of course also be interpreted in favour of the rotational motion of the HCl dipoles in the cubic lattice.

The phenomenon of molecular rotation in crystals probably does not occur quite so frequently as was supposed by Pauling, who suggested that the peculiar type of transformation exhibited by the other hydrogen halides, HBr and HI, and by a fairly large number of other substances, should be ascribed to this cause. These transformations differ from the ordinary allotropic crystal transformations like that of HCl (see fig. 8) or those of CO, $N_2$ and $O_2$, in that they do not take place at a

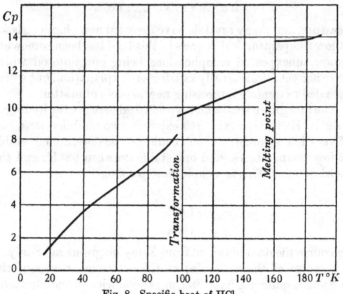

Fig. 8. Specific heat of HCl.

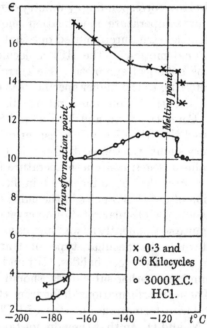

Fig. 9. Dielectric constant of HCl.

⟨ 120 ⟩

definite temperature. The absorption of heat accompanying the new type of transformation seems to begin at absolute zero, but is so small at low temperatures as to be hardly discernible. It gradually increases with temperature and finally rises to an extraordinarily high maximum (see fig. 10); at the temperature of this maximum the transformation is apparently finished and the heat effect abruptly drops again to

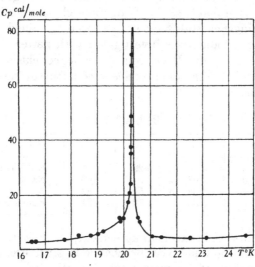

Fig. 10. Typical λ-point.

insignificantly small values. The shape of the specific heat curve resembles the Greek letter λ and, following Ehrenfest's suggestion, these transformations are referred to as λ-points.

The change in crystal structure with which we are concerned in this paragraph, if it exists at all, is extremely slight in all instances of λ-points. It easily escapes notice on the X-ray diagram, if one does not apply methods of high precision. In fact in many instances the change in crystal structure was established by other means.

Hydrogen bromide has no less than three λ-points with maxima at 89·2°, 113·4° and 116·8° abs. X-ray investigations have been reported by Natta and by Ruhemann. Natta's diagram, taken at 100° abs., showed lines which could be

explained by the assumption of a cubic cell, but none of the
diagrams of Ruhemann, taken below, between and above the
λ-temperatures, could be brought into accord with a regular
structure. No satisfactory interpretation of these diagrams
could be given, but in spite of some characteristic similarities
they are certainly different from each other, the one taken at
the lowest temperature being richest in lines. Again the
differences are not very marked, so that the change in crystal
structure is probably slight. Unfortunately this is all that can
be said; single crystals have not yet been examined.

Hydrogen iodide was investigated with better success by
the same authors. Both come to the conclusion that it
crystallises in a tetragonal face-centred lattice of the following
dimensions:

Table XIV

| $a$ | $c/a$ | $\rho$ | $T°$ K. | Observer |
|------|-------|--------|---------|----------|
| 6·19 | 1·08 | 3·29 | 125 | Ruhemann |
| 6·10 | 1·08 | . | 100 | Natta |
| 6·04 | 1·08 | 3·56 | 82 | Ruhemann |
| 5·95 | 1·08 | 3·72 | 21 | Ruhemann |

Their results are in excellent agreement, but it is striking that
the X-ray diagrams reveal no change in structure, although
two λ-transformations occur in the investigated temperature
region at 70·1° and 125·6° abs. It must be concluded that if
these λ-points are at all accompanied by such changes, they
must be either beyond the limits of accuracy of these investigations or of a kind which only affects the position of the hydrogen
nuclei, which, of course, cannot be located by X-rays.

That the processes giving rise to the λ-peaks in HBr and in
HI have something fundamental in common seems nevertheless demonstrated by the experiments of Smythe and Hitchcock. Fig. 11 $a$ and $b$ show the variation of the dielectric
constant with temperature for both substances; the similarity
of these curves is just as striking as their dissimilarity from
that for HCl (fig. 9). The behaviour at the lowest λ-points
rather recalls that of a "Seignettelectric" at its electric Curie
point.

So far the experimental material does not allow a definite conclusion to be made as to a possible connection between the λ-phenomenon and molecular rotation in HI and HBr

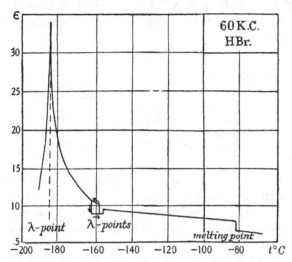

Fig. 11*a*. Dielectric constant of HBr.

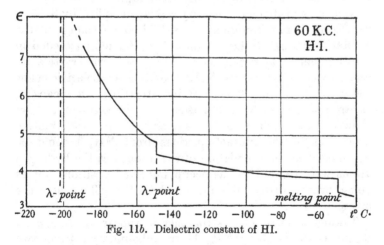

Fig. 11*b*. Dielectric constant of HI.

crystals, nor is the case very much clearer with respect to other substances with λ-points.

The classical examples are the *ammonium halides*. In

$NH_4Cl$ the phenomenon was discovered first (by Simon) and therefore they are sometimes referred to as transformations of the $NH_4Cl$ type. Table XV shows that the $\lambda$-temperatures for $NH_4Cl$, $NH_4Br$ and $NH_4I$ lie fairly close to each other, closer than their respective allotropic transitions, and vary in the same sense:

## Table XV

| Substance | $\lambda$-point ° K. | Allotropic transitions ° C. |
|---|---|---|
| Chloride | 242·8 | 184·3 |
| Bromide | 235·2 | 137·8 |
| Iodide | 230·7 | −17·6 |

The well-known lattice of the ammonium halides takes an intermediate place between ionic and molecular lattices; the high-temperature modification is of the $NaCl$ type, the one in which the $\lambda$-transitions occur of the $CsCl$ type, where the place of the alkali ion is occupied by the ammonium radical. It is, however, impossible to arrange the hydrogen atoms in such a way that a lattice of the crystallographic symmetry $O$ is obtained, even if the elementary cell is multiplied by 8, 27 or 64. Early X-ray investigations exhibited no change in the crystal structure at the $\lambda$-points. But Hettich succeeded in showing that, whereas at room temperature the crystals are certainly not piezoelectric, $NH_4Cl$ becomes piezoelectric below the $\lambda$-temperature. Structural investigations by means of electron rays carried out by Laschkarew at room temperature showed that slightly better agreement with the relative intensities of the diffraction rings from $NH_4Cl$ could be obtained if in the calculation of the intensities the hydrogen atoms were treated not as occupying fixed positions but as being evenly distributed on a spherical shell around the nitrogen atom.

Taking all these results together it appears as though in some way the $\lambda$-point of $NH_4Cl$ forms a boundary between two crystal modifications. One has a polar axis and can be pictured as in fig. 12 with the hydrogen atoms arranged on the trigonal axis of the cube, forming tetrahedrons around the nitrogen atoms that point in the direction of the polar axis. The other has $O$-symmetry, which is obtained through a random distribution or rotational motion of the hydrogen atoms about the nitrogen atom.

Ewald and Hermann give in the "Strukturbericht" another possible explanation of the $O$-symmetry of the crystals stable above the $\lambda$-point; it implies non-equivalent positions of the halogen ion. The edge of the elementary cube is doubled and the cell contains eight halogen ions, forming two face-centred lattices with the origins in $(0\,0\,0)$, $(\frac{1}{2}\frac{1}{2}\frac{1}{2})$, eight nitrogen

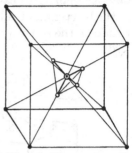

Fig. 12. Elementary cell of NH₄Cl lattice.

atoms also arranged in two face-centred lattices with the origins in $(\frac{1}{4}\frac{1}{4}\frac{1}{4})$, $(\frac{3}{4}\frac{3}{4}\frac{3}{4})$, and the hydrogen atoms with parameters $(\pm u \pm u \pm u)$. This explanation possibly holds for the NH₄Br lattice. In contrast to the chloride it does not become piezoelectric, but Hettich found it to show double refraction below the $\lambda$-point. According to Ketelaar, in the Debye diagrams taken at $-100°$ C. faint extra lines appear. These indicate a tetragonal cell with $a$ equal to $a'\sqrt{2}$, if $a'$ denotes the edge of the original cube (half of that proposed by Hermann) and $c = a'$, containing two molecules. From intensity considerations follows the space group $O_{4h}^7$; the nitrogen atoms have the co-ordinates $(0\,0\,\frac{1}{2})$, $(\frac{1}{2}\frac{1}{2}\frac{1}{2})$ and the bromine atoms $(0\,\frac{1}{2}u)$, $(\frac{1}{2}\,0\,\bar{u})$. The parameter $u$ is equal to $0 \cdot 030 \pm 0 \cdot 005$. Ketelaar found the same behaviour for the iodide; in the latter case the parameter is much smaller: $u = 0 \cdot 01 \pm 0 \cdot 005$, and accordingly the extra lines are much fainter. Fig. 13 illustrates the connection between the cubic lattice of Hermann and the tetragonal lattice of Ketelaar. The arrows show how the bromine atoms occupying non-equivalent positions according to Hermann are displaced by the parameter $u$ in opposite directions from their original positions to those in the tetragonal lattice.

It appears again that just as in the case of chloride, in the bromide and the iodide the $\lambda$-temperature separates two different crystal modifications. However, the details are markedly different and a connection between the λ-point and the beginning of rotation of the ammonium radical is much less obvious than in the case of chloride. An apparent dissimilarity in the behaviour of chloride on the one hand and bromide and iodide on the other we have already noticed in the case of the hydrogen halides. A glance back at Table XV, where we see

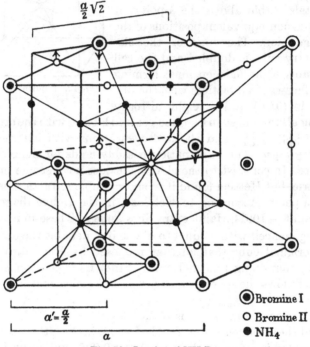

Fig. 13. Lattice of NH₄Br.

how small the influence of the halogen ion on the λ-temperature is, as compared with that on the transition temperature from the NaCl type lattice to the CsCl type lattice, warns us that the dissimilarity in the case of the ammonium halides will be rather difficult to interpret.

We encounter the same difficulty with two other analogously built substances: $CH_4$ and $SiH_4$. In both λ-points have been observed through specific heat determinations. $SiH_4$ was proved by Clusius to undergo a change in crystal structure.

He used Wahl's method of observing the crystals formed in a thin-walled quartz vessel immersed in a cooling liquid under a polarisation microscope and found that the weak double refraction observed above the λ-peak (65° K.) changes below the peak, where new crystals are formed with very strong double refraction. There are, unfortunately, no X-ray data for $SiH_4$, but the structural change seems to be established beyond doubt.

Now methane has been examined very carefully by a number of authors, but no sign of a change of crystal structure could be discovered, although the λ-phenomenon is very marked (see fig. 10.) The specific heat reaches a maximum value of 81 calories per degree at 20·4° K. Wahl had reported in 1914 that a doubly refracting modification could be observed below 20° K., but a repetition of his experiment by Clusius showed that no transformation takes place and that the crystals are cubic from 15° K. to the melting point, although at all temperatures they may sometimes show slight double refraction due to internal strains. X-ray investigations led to the same result. Mooy took Debye diagrams of methane at various temperatures above and below 20·4° K. and found them to be perfectly identical both as regards spacing and intensity. He found the lattice to be cubic, the C-atoms forming a face-centred cell. The structure is, however, not a close packing, as McLennan had assumed; a careful estimation of intensities showed that they were markedly different from those of a close packing, thus affording the possibility of determining the positions of the H-atoms. On the assumption of crystallographically equivalent positions for the four C-atoms and for the sixteen H-atoms as well, only two space groups are possible, $T$ and $T_d^2$; in both the co-ordinates of the C-atoms are as mentioned above, $(0\,0\,0)$, $(\frac{1}{2}\frac{1}{2}0)$, $(\frac{1}{2}0\frac{1}{2})$, $(0\frac{1}{2}\frac{1}{2})$, and those of the H-atoms $(u\,u\,u)$, $(u\,\bar{u}\,\bar{u})$, $(\bar{u}\,u\,\bar{u})$, $(\bar{u}\,\bar{u}\,u)$, and so on. By comparing the measured intensities with those calculated for both these space groups and for different values of the parameter, Mooy succeeded in proving that the space group is $T_d^2$ and $u$ is approximately equal to 1 Å. His result is very striking and rather unexpected. It was generally accepted that above the λ-point the H-atoms were rotating about the C-atoms, in which case it should be impossible to assign them definite positions in the lattice and the space group should not be a $T$ but an $O$ group. However, the effect in the intensities found by Mooy appears to be real, and that excludes the assumption of molecular rotation as an explanation for the $\lambda$-phenomenon in methane. This is a very unsatisfactory result, as, in view of the absence of a change in crystal structure, no other explanation is at hand.

Of other organic compounds that have been investigated at low temperatures we shall refer here only to *ethylene*. This substance was investigated by Keesom in the form of single crystals by the method described above. By means of a specially devised gnomonic projection he found three reflections from mutual perpendicular crystal planes and concluded that the symmetry should at least be rhombic. Debye diagrams obtained formerly by Mooy had shown that the symmetry is at most rhombic. Thus the indexing of the diagram was carried out for a rhombic cell, for which the following dimensions were found:

Table XVI

| $a$ | $b$ | $c$ | $a/b$ | $c/b$ | $\rho$ | $T$ |
|------|------|------|-------|-------|--------|--------|
| 6·46 | 4·87 | 4·14 | 1·327 | 0·850 | 0·717 | 98° K. |

The density has been measured only at 20° K. and in the liquid $\rho_{20^\circ} = 0 \cdot 78$, $\rho_{120^\circ} = 0 \cdot 62$. The density given in Table XVI was calculated for the case of two molecules $C_2H_4$ in the cell.

On the assumption that each $CH_2$ group acts as one centre, an arrangement of these goups with the co-ordinates $(0\cdot16, 0, 0)$; $(\frac{1}{2}, \frac{1}{2}, \frac{1}{2}+0\cdot12)$; $(-0\cdot16, 0, 0)$; $(\frac{1}{2}, \frac{1}{2}, \frac{1}{2}-0\cdot12)$ leads to a satisfactory interpretation of the intensities. The H-atoms could not be located in this structure, as the results of the intensity calculation were not influenced even by the assumption of a very unfavourable distribution of the H-atoms.

Methane is one of the very few cases where the position of the H-atoms can be determined by X-rays. This possibility depends largely on the relative atomic numbers of the various atoms in the compound and on the character of binding; e.g. in LiH, where hydrogen appears as a negative ion with two electrons, lithium being very light, the position of the hydrogen ion could be successfully determined. In compounds where hydrogen is a positive ion this is impossible and in non-polar compounds at any rate very difficult. Vegard in his discussion of the structure of $H_2S$ and $H_2Se$, on the ground of experimental material for many crystals and mixed crystals containing hydrogen, comes to the conclusion that in non-polar compounds it should be possible to ascribe definite positions to the H-atoms from crystallographic considerations. For the two analogous substances $H_2S$ and $H_2Se$ he finds that the heavier atoms or the molecular centres are arranged in a cubic face-centred lattice, and that consideration of the H-atoms leads to the space group $T_h^6$. This refers to the modification stable at liquid-air temperature. The modification of $H_2S$ above $103.5°$ K. has not been examined, and one cannot a priori exclude the possibility that the H-atoms in this upper modification do not occupy fixed positions but rotate in the lattice. In such cases it is always desirable to supplement the X-ray analysis by other physical methods of examination, as e.g. the absorption and Raman spectra, dielectric constants, etc.

In the very interesting case of $N_2H_4$ the decision between the two possible structures proposed by Vegard from Debye diagrams was also not obtained by X-rays. The reason is here very different; although the symmetry is cubic, the diagrams contain so many lines that not less than six molecules of $N_2O_4$ or twelve $NO_2$ molecules must be assumed in the elementary cell. The discussion of the intensity is therefore very difficult. Vegard succeeded in reducing the number of possible space groups to two, $T^3$ and $T^5$; $T^3$ is compatible with $N_2O_4$ molecules, $T^5$ with $NO_2$ molecules. Now in the gaseous and liquid states an equilibrium exists between $NO_2$ and $N_2O_4$ which with falling temperature varies in favour of $N_2O_4$; the existence of $N_2O_4$ molecules in the solid state was proved from the Raman spectrum of the crystal; therefore the space group $T^3$ must be considered as correct.

The question of association in the solid state is also very important for the crystals of NO, particularly with respect to its magnetic susceptibility. Unfortunately crystallographic data have not yet been obtained.

We have here given an account of such crystal structures

as have been investigated at low temperatures, mainly using material that has been published after the publication of the "Strukturbericht". Apart from that we have paid particular attention to such investigations which throw light on physically interesting questions. Of these one comes to the forefront, which is indeed so important that we shall discuss it more fully in the subsequent paragraph: the question of molecular rotation in crystals.

## II. i. 3. *Molecular Rotation in Crystals*

The question whether molecules are free to rotate in a crystal lattice was raised in a general manner by Pauling. By an argument much resembling that with which Lindemann derived his melting point formula, he showed that if such molecules pick up rotational energy when heated from absolute zero to the melting point, a transition will in some cases occur. At absolute zero the molecules are supposed to perform a librational motion about the equilibrium position; if they succeed in acquiring a sufficient amount of librational energy before the solid melts, this oscillatory motion should change into rotational motion in the solid at a limiting temperature. This temperature can be roughly estimated by quantum theoretical considerations from the heat capacity of the solid, if this can be sufficiently accurately separated into the part contributed by the ordinary lattice vibrations and that due to the librational motion. In practice this is rarely the case; but Pauling suggested that the $\lambda$-anomalies might be connected with the phenomenon of molecular rotation. The connection is not very obvious, since rotational specific heats should in fact be very different from the characteristic $\lambda$-shape of these anomalies, and all authors agree that at least additional hypotheses have to be introduced. Moreover, from the foregoing chapter we inferred that it is by no means easy to establish experimentally whether molecules are rotating in a crystal or not, and the evidence for this in the case of the $\lambda$-anomalies is very contradictory. In the present state of affairs it thus appears advisable to treat the two problems independently.

The problem of molecular rotation in crystals has been treated theoretically in great detail; it is a matter of finding the eigenfunctions of a molecule in a field in which it is capable of occupying two or more positions of equilibrium. This is fairly easy in two dimensions, but for the case of the three-dimensional crystal field it involves rather elaborate quantum mechanical calculations, which are of considerable theoretical interest. These calculations have so far not been carried to a point which allows accurate testing by experimental data. Therefore we will here give merely a short survey of such cases in which experimental evidence exists for or against a rotation effect and which may perhaps in future serve as a basis for the application of the theory.

The main methods by which rotational motion of molecules in a crystal can be established are the following:

(1) Crystal-structure analysis by means of X-rays or electron rays. The evidence is obtained from the intensities of the deflected rays, which in case of rotation will not, as is usually the case, allow the determination of fixed positions for the atoms of the rotating group. The evidence is inconclusive if the group is made up of atoms of very different atomic weight, as the intensities are not sufficiently influenced by the reflections from the light atoms.

(2) Piezoelectric experiments. If a crystal is piezoelectric below a given temperature and loses this property above, this indicates that atoms determining a polar axis at low temperatures change their position in the lattice. Together with a careful consideration of the crystallographical symmetry properties of the lattice this may lead to the conclusion that the molecules are rotating.

(3) The appearance of or a considerable change in the double-refracting power indicates a change of crystal structure which can in some instances be ascribed to the setting in of molecular rotation.

(4) Measurements of the dielectric constant. A negligibly small value of $\epsilon$ indicates a lack of rotational freedom of the dipoles; $\epsilon$ ordinarily rises sharply at the melting point, where this rotational freedom is gained. If this sharp and

considerable rise is observed at a temperature below the melting point, and if at the same time the change in $\epsilon$ at the melting point is small, it may be concluded that the dipoles gain rotational freedom already in the solid state.

(5) Optical measurements. Observation of the optical absorption spectrum of a crystal, especially in the infra-red region, can lead to the determination of the rotational frequency and the moment of inertia of the molecule in the crystal lattice. If these are found, it is possible to calculate the effect of rotation on the specific heat. Moreover, such determinations can provide the quantities needed for the theoretical calculations. Optical measurements are therefore extremely important in this respect.

To investigations of the Raman spectrum of crystals the above remarks also apply. Not in all cases could the absorption spectrum be successfully interpreted. But the bands mostly show a considerable temperature-dependent displacement, which is abnormally large in the neighbourhood of a transition.

It must be emphasised that the interpretation of results obtained from these methods is not always conclusive, and a final answer to the question can be obtained only by combined consideration of all of them.

In order to give a comprehensive survey of the very bulky and non-homogeneous experimental material we have summarised it in Table XVII. In the first column are listed a number of substances which have been investigated with a view to our question. Unless the whole molecule is supposed to be rotating, that part which is has been emphasised by brackets. In the second column are listed the temperatures of any transitions that may occur in the substance and in the next whether it is a transition of the ordinary type (ord.) or a $\lambda$-point ($\lambda$). If other transitions occur, which are not directly of interest with respect to our problem, this is indicated in brackets. Then follow the various methods by which evidence was sought; the sign " + " means that the evidence is definitely in favour of rotation; "0" means that an effect has been

Table XVII. *Molecular rotation in crystals*

| Substance | Transition temperature | Type of transition | X-ray evidence | Piezo-electric effect | Double refraction | Dielectric constant | Rotational absorption or Raman lines | Temperature effect on absorption spectrum | Interpretation of specific heat | Conclusional remarks |
|---|---|---|---|---|---|---|---|---|---|---|
| Hydrogen | · | none | + | · | 0 | · | · | · | + | Probably rotating down to absolute zero |
| Oxygen | 23·5 43·8 | Ord. Ord. | + | · | 0 | · | · | 0 | − | Evidence conflicting |
| Nitrogen | 35·5 | Ord. | ++ | · | 0 | · | · | · | − | Evidence conflicting |
| CO | 61·5 | Ord. | ++ | · | 0 | · | · | · | − | Evidence conflicting |
| HCl | 98·3 | Ord. | · | · | · | + | 3·7$\mu$ | 0 | +? | Rotation in upper modification very probable |
| HBr | 89·2 113·4 116·8 | λ | 0 | · | · | 0 | 0 | · | ? | No definite evidence |
| HF | 70·1 | λ | 0 | + | No | 0 | 0 | · | ? | No definite evidence |
| (NH$_4$)Cl | 125·0 242·8 | λ (ord.) | No; electron rays + | No | · | · | + | 0 | ? | Rotation above 242·8° K. very probable |
| (NH$_4$)Br | 235·2 | λ (ord.) | 0 | No | 0 | · | 0 | 0 | ? | No definite evidence |
| (NH$_4$)F | 230·7 | λ (ord.) | No | · | 0 | · | 0 | 0 | ? | No definite evidence |
| (NH$_4$)(NO$_3$) | 212·0 +125·2° C. | λ (3 ord.) Ord. | (0) + | No | · + | · − | + | 0 | ? ? | NO$_3$ rotates above 125° C. NH$_4$ probably above 272° K. |
| NH$_3$ | | No | + | · | · | · | · | · | ? | No rotation |
| CH$_4$ | 20·4 | λ | − | · | − | − | − | · | ? | No rotation? |
| SiH$_4$ | 63·5 | λ | · | · | 0 | · | 0 | · | · | No definite evidence |
| H$_2$S | 103·5 126·5 | Ord. Ord. | ·0 | · | · | + | · | · | · | Probably rotating above 103·5° C. |
| Methanol | 160·0 | λ | ·+ | · | ·0 | + | · | · | ? | Probably rotating above 160° C. |
| Na(NO$_3$) | +275·5° C. | λ (ord.) | · | · | · | · | · | · | · | Probably rotating above 275·5° C. |

observed but does not necessarily lead to the assumption of rotating molecules. "No" indicates that the effect which was looked for has not been found, but that it does not follow that the molecules do not rotate. " − " is used if the experimental result is incompatible with the assumption of rotation. Under the heading "Interpretation of specific heat" we find many question marks, indicating that an exact quantitative interpretation in terms of rotational frequencies has not yet been found possible.

We conclude from this table that in seven cases the experimental evidence points more or less strongly to a rotational motion of molecules or groups in the crystalline state: $H_2$ is probably rotating from absolute zero; HCl, the $NO_3$ group in ammonium nitrate and $H_2S$ apparently start rotating above an ordinary transition point, and the $NH_4$ group in ammonium chloride and ammonium nitrate, the $NO_3$ group in sodium nitrate and probably the hydroxyl group of methanol (if not the whole methanol molecule) commence rotation above a λ-point. A quantitative analysis of the specific heat (not the heat of transition) in terms of rotational frequencies has, however, been attempted only for HCl. It depends on the interpretation of the double band at $3 \cdot 7\mu$ found by Hettner; Callihan and Salant conclude from Raman data that the two parts of the band should be interpreted as a $Q$- and a $P$-branch, and thus remove the difficulties encountered by Hettner, who from his own interpretation obtained much too large rotational specific heats. Apart from the difficulty which the authors mention, i.e. that in this case the $R$-branch would inexplicably be missing, there are other points of doubt: e.g. the Raman data refer to the upper modification, whereas the double band was observed by Hettner only for the lower modification, where it is subject to a complicated change with temperature. The upper modification exhibits in the infra-red only a single band, which is almost identical with that of the liquid; and it is this upper modification in which we should expect the molecules to be rotating freely. This instance demonstrates what difficulties and complications we encounter, if we wish to test the idea of molecular rotation in crystals, and what an enormous amount of work has yet to be done even in this case, which is certainly the most thoroughly investigated, before we can treat it as an established fact and apply the theoretical calculations referred to above. The problem as a whole is certainly one of the most interesting tasks eventually to be solved mainly through experiments at low temperatures.

# CHAPTER II

## THE THERMAL ENERGY OF CRYSTALS

### II. II. 1. *Low Temperature Calorimetry*

The discussions of the foregoing chapter led us to recognise the importance of an accurate knowledge of specific heats. Indeed, in the course of his research the low-temperature physicist is so often forced to acknowledge the extraordinary wealth of information on the solid state that can be obtained from such measurements that one is tempted to attribute to them more fundamental significance than to any other method of investigation. With respect to high temperatures (1000° C.) such a statement would appear rather exaggerated; even at normal temperatures it is hardly correct. But the deeper we proceed into the region of low temperatures, the nearer it comes to the truth. This is partly due to the high sensitivity of low temperature calorimetry, which increases with diminishing temperature.

The instrument which enables these accurate measurements to be carried out is Nernst's vacuum calorimeter. It is an instrument most superbly suited for its purpose and it may well be said that low-temperature physics dates from its invention. It is to low-temperature research what the spectrograph is to spectroscopy and the Wilson chamber or the Geiger counter to nuclear physics.

Extremely simple in principle, it has been developed into a rather complicated apparatus for special purposes. The oldest and most primitive form simply consists of a block of the material to be investigated, suspended in a vacuum vessel (fig. 1). Around the block is wound a fine wire, through which electrical energy in measured quantities is applied to the specimen and by the resistance of which the temperature rise produced in the specimen can be determined. The whole is surrounded by a cooling liquid. Hydrogen or helium gas is introduced into the vacuum vessel to bring the specimen to the temperature of the liquid, and is afterwards pumped off for adiabatic insulation. This is in short the original apparatus used by Nernst and Eucken for the first determinations of specific heats at temperatures down to those of liquid hydrogen.

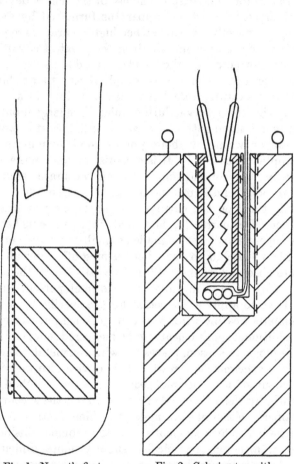

Fig. 1. Nernst's first        Fig. 2. Calorimeter with
vacuum calorimeter.                heating core.

This simple calorimeter allows the measurement of specific heats of metals with an accuracy of about 2 per cent.

The improvements that have been introduced are of two kinds: they concern either the technique of the measurements or the construction of the calorimeter.

The measurement of the energy input involves the determination of voltage, current and time; that of temperature requires the same electrical data and an accurate standard of temperature for comparison. The use of the highly developed but rather costly electrical apparatus furnished by modern industry is certainly advisable when high accuracy is required, but gives a noticeable effect only if the greatest attention is paid to the correctness of the standards used to determine time and temperature. The efforts of physicists to establish a correct temperature scale have been described in a former chapter. The time interval during which the energy input lasts is usually measured with a stop-watch: ordinary stop-watches are, however, wrong by about $\frac{1}{5}$ of a second per minute, very good stop-watches by $\frac{1}{50}$. It is essential to control the watch repeatedly by comparison with an accurate standard, e.g. a second pendulum, and to use a switch which automatically turns on the current and starts the watch at the same time. More correctly, the period of time elapsing between the two operations must be made as small as possible and the same when switching on as when turning off. After all these rules are observed we may still not be satisfied by the results and we may have to improve the calorimeter itself. First of all we have to apply a correction for the specific heat of the thermometer-heating wire. This can be done more accurately if the additional specific heat can be measured independently. For this purpose Keesom places the wire in a metal "core", to which the specimen is screwed. This device also makes the replacement of the specimen easier. Another method used frequently by Nernst's collaborators is to put the specimen in a container of either glass or metal, to which the thermometer is attached. The arrangement of the thermometer inside this container, although technically slightly more difficult, is advisable, as it ensures better heat contact with the specimen and decreases the correction for heat losses from the thermometer due to radiation.

Good heat contact between specimen and thermometer even in the case of substances of small heat conductivity, e.g. powders or solidified gases, can be secured by introducing a non-sorbable gas, $H_2$ or He, at small pressure or by inserting vanes of thin copper. The use of such devices is limited only by the requirement that the heat capacity of the calorimeter should be kept as small as possible in comparison with that of the specimen.

To shield the calorimeter from heat radiation as well as conduction it is best to surround it by a special jacket, which by means of a separate heating wire is kept continually very close to the temperature of the calorimeter. The temperature difference is controlled by a thermocouple. All leads to the calorimeter are first brought into good heat contact with this shield, which is made of heavy material with high heat capacity and weighted with lead. This also serves as a buffer between the former and the surrounding temperature bath against heat influx through conduction.

A very important variation of this adiabatic shield is the so-called condensation ring. The place of the lead block is taken by a ring-shaped vessel into which a cooling liquid can be condensed, e.g. hydrogen. The shield with ring is made to enclose the calorimeter vacuum-tight. When the whole apparatus has assumed the temperature of the surrounding hydrogen bath, a quantity of hydrogen is condensed into this ring. Then the contact gas is pumped off from the outer container, but not from the space between the shield and the calorimeter. The vapour pressure above the liquid hydrogen is reduced by means of a powerful vacuum pump and thus the temperature of shield and calorimeter is very efficiently lowered. When the temperature has been reduced as far as possible, the shield is also evacuated and the measurements can begin. The heat of evaporation of the liquid hydrogen remaining in the ring serves the same purpose as the heat capacity of the lead block described above.

The ring calorimeter can of course be used in other temperature regions also, say with nitrogen or helium. Its special advantage is the possibility of lowering the temperature of the calorimeter considerably below that of the bath. It thus marks an important step towards filling the gap between the temperature regions covered by the different cooling liquids. To pump off the liquid in the outer Dewar vessel to 12° K. in the case of hydrogen and 60° K. in the case of nitrogen a

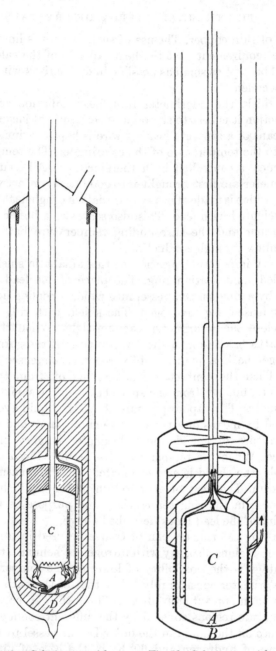

Fig. 3a. Calorimeter with
adiabatic shield.

Fig. 3b. Calorimeter with
condensation ring.

powerful pump is needed; whereas with the well-shielded ring we can easily obtain 9° K. or 50° K. respectively with the help of an ordinary rotating oil pump.

All the elements of construction that we have here described have been used in various combinations according to the special purpose of the individual investigation. For further details we refer to the original papers listed at the end of the book. We may add a few remarks concerning the choice of the most suitable material for a calorimeter. It is dictated by two requirements: good thermal contact inside the calorimeter and good thermal insulation from the surroundings. Thus a combination of glass and metal seems the happiest solution: metal for the calorimeter, glass for the tubing. This solution was chosen by Giauque. Eucken and Clusius have found all-glass calorimeters useful. Simon and his collaborators have preferred all-metal ones, using German silver for the tubing. The suitability of various materials can approximately be judged from the following figures:

| | Glass | German silver | Brass | Copper |
|---|---|---|---|---|
| $\lambda$ | 1 | 5 | 25 | 125 |

where $\lambda$ designates the relative amount of heat conducted by tubes made of these materials in unit time between two given temperatures, if the tubes are of equal length, of 8 mm. inner diameter and a wall strength guaranteeing sufficient rigidity of the tubing.

## II. II. 2.   *The Specific Heats of Crystals*

Debye's theory of specific heats has shared the common fate of all theories. The results of the first ten years of low-temperature calorimetry culminate in the fact that a good agreement between theory and experiment is recorded. The rich material contributed mainly by Nernst and his collaborators confirmed the general form of the specific-heat curve prescribed by Debye's interpolation formula, as well as the $T^3$ law within the limits of accuracy then obtainable, very convincingly. In the following period it was again Nernst's school that took the lead in the development, which eventually showed that Debye's assumptions are much too primitive to account for the complicated phenomena observed even in such simple cases as regularly crystallising elements. In itself this discovery is of course anything but a misfortune. It is, however, most disappointing that we are not in a position to offer a satisfactory alternative. In order to be able to explain deviations from an expected course we must first, on firm grounds, know what course to expect. Unfortunately there seems to be no theoretical means to gauge the degree of accuracy with which the Debye function can reasonably be expected to hold. The improvements introduced by Born and Kármán lead to a function which deviates very little from Debye's; but Blackman, who has further developed the theory on the lines of Born, expresses the opinion that whereas formerly the Nernst-Lindemann formula was considered to be a good first approximation to the Debye formula, nowadays the latter should rather be considered a first approximation to the former. However this may be, it seems rather evident that the final solution will have to be connected with a complete quantum-mechanical treatment of the crystal lattice, which may not be achieved for a long time.

Debye's theory deals exclusively with that part of the energy content of a crystal which is due to the thermal vibrations of the atoms about their equilibrium positions. To this we shall in future refer as the normal specific heat, in the case of molecular lattices including in this the vibrations of the molecular centres about their equilibrium position as well as the vibrations of the atoms of one molecule with respect to each other. There are, however, other causes of energy consumption theoretically conceivable. Of these we have already mentioned in a former chapter the rotational motion of the molecules. Another contribution is made by the conductivity electrons of a metal. A third is the possible distribution of atoms on different energy levels, which are successively excited with rising temperature. Somewhat analogous is the proposition that the atoms may be oriented to the crystal axis in different ways, these different atoms being arranged in perfect order at absolute zero but disorder ensuing with rising temperature. Before we enter into the discussion of these anomalies we must emphasise that owing to the uncertainty in the theoretical course of the normal specific heat it is in practice often very difficult to analyse the experimental curves with a view to establishing the connection with any one of these theoretically possible heat effects.

The anomalies observed in the specific heats of a great number of substances in different regions of low temperatures are of an extraordinary variety; for the purposes of description we shall try to group them roughly in a phenomenological way.

On analysing the specific-heat curve according to Debye's theory it has frequently been found that a single $\Theta_D$ is not capable of describing the whole course of the measured specific heat. These deviations generally become noticeable only at rather low temperatures in and below the hydrogen region, and are conveniently demonstrated in a graph showing the variation of $\Theta_D$ with the temperature (fig. 4). As is seen in fig. 4, these deviations are by no means regular and no general rule can be given; for some substances $\Theta_D$ shows a maximum, for others it varies more smoothly; but neither the steepness of variation nor its direction is uniform. This lack of uniformity makes it difficult to attribute these deviations to a particular physical phenomenon; partly they must rather be taken as a sign of the general inadequacy of Debye's theory. As is well known, Debye's model for the energy spectrum of a crystal is the same as for a continuum, only broken off at a plausible frequency. It is not astonishing that this rather rough model cannot at all temperatures adequately describe the conditions in a crystal. It was, however, thought that at temperatures below about $\frac{1}{10}\Theta_D$ the $T^3$ region should begin, i.e. all crystals should have a specific heat proportional to the cube of the temperature. This law is the limiting case for very low temperatures and does not depend on the special features of the Debye model. We should therefore expect it to be valid even for non-regular crystals. Now we see in the graph (fig. 4) that a $T^3$ region is indeed found in almost all instances; but where measurements have extended to temperatures of about $\frac{1}{20}$ of $\Theta_D$ or lower all metals have been shown to possess a

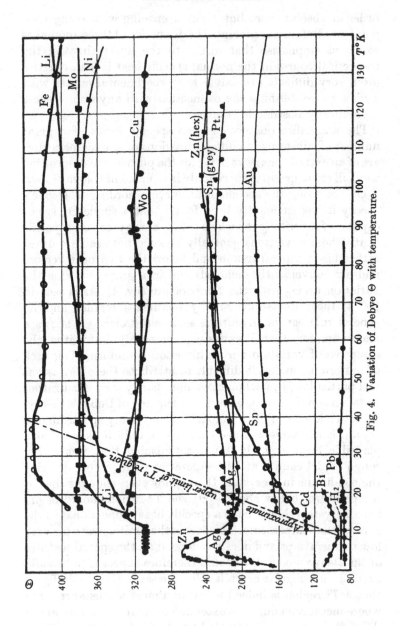

Fig. 4. Variation of Debye Θ with temperature.

considerably larger specific heat than corresponds to the extrapolation of the $T^3$ law. This is expressed in the sharp drop of the $\Theta_D$-values for silver, zinc, iron, nickel and beryllium. In the case of nickel this drop is so colossal that there is no sense

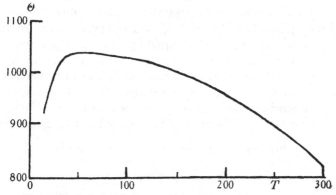

Fig. 5. Variation of Debye $\Theta$ of beryllium with temperature.

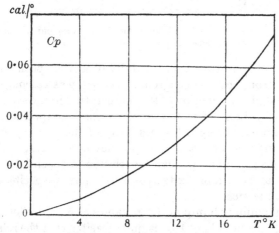

Fig. 6. Atomic heat of nickel.

in applying the theory at all; at 2° K. the specific heat is about forty-five times that of the extrapolated Debye curve, changing linearly with the temperature (see fig. 6). It is, of course, very tempting to attribute this additional specific heat to the free electrons in the metal, and many investigations have been carried out with a view to finding this effect. At first these attempts did not lead to very definite results, since the variability of $\Theta_D$ makes it impossible to give even an approximate estimation of the normal specific heat. Evidently one should study the behaviour of a non-metallic substance for comparison. KCl is very suitable for this purpose; the atomic weights of K and Cl are nearly equal, the lattice is of the NaCl type and, moreover, down to the temperatures of liquid hydrogen its $\Theta_D$ is constant and equals 220, i.e. almost equal to that of silver. The measurements carried out by Keesom and Clark lead to a very satisfactory result. The comparison between KCl and Ag was made in the following manner: the specific heat of the free electrons as calculated from Sommerfeld's theory for 1 el./atom were

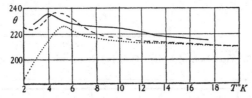

Fig. 7. Debye $\Theta$ for KCl (full line), Ag (dotted line) and Ag corrected for electron specific heat (broken line).

subtracted from the experimentally observed specific heat of silver. From the resulting curve the $\Theta_D$ was calculated and represented as a function of the temperature in the same graph with the values derived from the measured specific heat of KCl. As we see in fig. 7 the similarity of the two $\Theta_D$-curves is so striking that one cannot help believing that a specific-heat curve corresponding to such a variation in $\Theta_D$ should be considered as fairly representative for the normal specific heat of a crystal lattice.

The remaining hump in the $\Theta$-values at 5° K. indicates that the $T^3$ law is not obeyed. It is quite feasible that the relatively good constancy of $\Theta$ in the beginning of the $T^3$ region (see fig. 4) is not really of theoretical significance. Blackman has pointed out that this need not really be expected at these temperatures. From his analysis of the energy spectrum of crystals he is led to the conclusion that the true $T^3$ region should be reached only at about $\frac{1}{50}\Theta_D$. Still, the significance

⟨ 146 ⟩

of the hump in the $\Theta$-curve is not yet clear, and in any case a definite opinion on the whole problem can only be formed after more experimental material is available.

If the coincidence of measured specific heats with the Debye curve is accidental in the $T^3$ region, there is no reason to suppose that it is less so at higher temperatures, where the Debye function has the character of an interpolation formula. And yet the impression we get from fig. 4 just in this region rather tends to show that as an interpolation formula it does very well indeed. We notice the excellent constancy of $\Theta$ for molybdenum, nickel, iron, hydrogen and lead over fairly large temperature regions; and where variations are observed such as in platinum, gold (zinc does not count because it is hexagonal) or lead they hardly exceed a few per cent. It appears then that a fairly normal behaviour of all these substances is guaranteed and that the Debye function is not so very inadequate in describing this normal behaviour.

There are a few noticeable exceptions, and it seems highly probable that to these exceptions more weight must be attributed than to the smaller deviations just mentioned. If abnormal specific heats are observed in the ferromagnetic metals Ni and Fe above 100° K. this is perhaps only natural, and a definitely abnormal behaviour of the coefficient of expansion confirms this (see fig. 8), nor will anybody be surprised by the discovery of a marked abnormal behaviour of Bi. These abnormalities are clearly expressed in the variation of $\Theta_D$, and we are thus led to the conclusion that a strikingly large discrepancy from the Debye formula indicates the presence of an abnormal specific heat which cannot easily be attributed to the imperfection of the theory of lattice vibrations. Our attention is clearly drawn to Li and grey Sn, the $\Theta$-curves of which cross those of several other substances, exhibiting a very marked abnormal increase of $\Theta$ with temperature. In contrast to Pt and Au, where we also noticed a certain increase in $\Theta$, the rise in the case of Li and Sn continues to very much higher temperatures; at room temperature Li has a $\Theta$ of 430, Sn 260. Thus the total variation between 15° K. and room temperature in the case of Li amounts to 33 per cent. and in that of Sn even to 57 per cent.

We have come to the second kind of anomalies, established experimentally, and we shall learn presently that they can be successfully interpreted by the third theoretical cause of abnormal energy consumption mentioned on p. 142. Schottky had pointed out that it should be conceivable that atoms of the same chemical nature might still be physically different; they might possess different energy, even in the solid state. At very

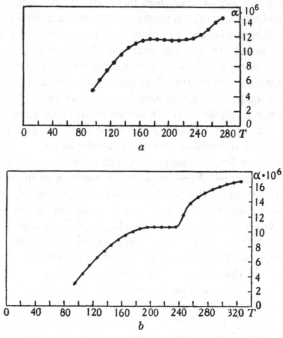

Fig. 8. Expansion of (a) nickel, (b) iron.

low temperatures, i.e. at temperatures very low compared with the difference in energy of the two kinds of atoms, all atoms will be of the kind with the lower energy. On heating, the energy input will be partly used for conveying to some of the atoms the energy necessary to change them into atoms of the other kind. The concentration of these atoms will thus depend on the temperature, as illustrated in fig. 9; it rises from 0 at absolute zero to 50 per cent. at very high temperatures, where atoms of both kinds will be available in equal numbers. In computing this relation the important assumption is made that the transition of one molecule is independent of the state of the rest, or that with respect to this transition the molecules behave like an ideal gas and do not influence each other.

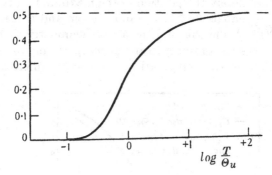

Fig. 9. Schottky function (concentration).

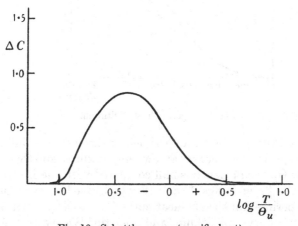

Fig. 10. Schottky curve (specific heat).

The calculation and theoretical implications of this conception will be discussed in a later chapter, so we shall concern ourselves here only with the comparison of the theoretical and experimental results. Now the heat effect due to this rise in concentration of high energy atoms is given by the curve in fig. 10. Both curves are represented as functions of the characteristic temperature of the anomaly $\Theta_U$, which is

⟨ 149 ⟩

introduced into the calculation as the energy difference of the two kinds of atoms per gramme-atom. We infer from figs. 9 and 10 that both curves are of universal character, neither their shape nor the maxima depend on $\Theta_U$, which latter determines merely the position of the maximum on the $\log T$ scale. $C_{max.}$ equals 0·87 cal./gramme-atom and is reached at $T = 0·42\Theta_U$. From this and the almost symmetrical shape of the $\Delta C$-curve with respect to an axis proportional to $\log T$ we must conclude that only rather low energy differences will

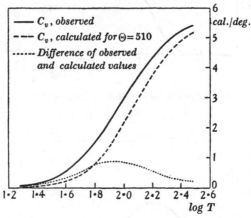

Fig. 11. Specific heat of lithium.

come to the notice of the experimenter through their influence on the specific-heat curves at low temperatures, and these only if their $\Theta_U$ is sufficiently small compared with the Debye $\Theta_D$. This is best seen in figs. 11 and 12, which give the separation of the experimental specific-heat curve into a Debye part and a Schottky curve for lithium and grey tin. We see that for Li with $\Theta_U = 0·39\Theta_D$ the anomalous behaviour is slightly less marked than for Sn with $\Theta_U = 0·26\Theta_D$. Simon has found that for a number of substances in the same groups of the periodic system with Li and Sn and having the same crystal structure the specific-heat curve in spite of a seemingly normal behaviour can be analysed in quite the same way, and that the new $\Theta_D$ determined in this way is in rather better agreement with the limiting frequency of the energy spectrum of these crystals as determined by methods independent of the specific heat. His results are summarised in Table XVIII:

### Table XVIII

| Substance | $\Theta_D$ sp. | $\Theta_D$ | $\Theta_D$ from other determinations | $\Theta_U$ | $U$ (volt-el.) |
|---|---|---|---|---|---|
| Li | 330–430 | 505 | 376 (el.), 500 (m.p.), 510 (exp.) | 205 | 0·0175 |
| Na | 159 | 202 | 202–208 (el.), 202 (m.p.) | 95 | 0·0082 |
| K | 99·5 | 126 | 115 (el.), 123 (m.p.) | 59 | 0·0051 |
| Diamond | 1840 | 2340 | . | 1070 | 0·092 |
| Si | . | 790 | . | 246 | 0·021 |
| Grey Sn | 140–260 | 260 | . | 69 | 0·0060 |

(el.) electrical conductivity; (m.p.) melting point;
(exp.) expansion coefficient.
sp. specific heat, directly determined.

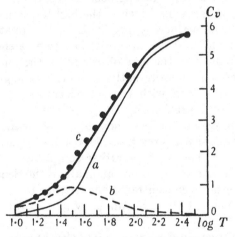

Fig. 12. Specific heat of grey tin.

The conclusion that Schottky anomalies have been proved in all these cases has often been criticised. It is held that this procedure shows more belief in the accuracy of the Debye theory than the latter deserves, and moreover destroys the very foundations on which this belief rests. For if we take substances that behave fairly normally and have thus been regarded as a proof of the theory, and show that this corroboration is fortuitous, where do we take the justification of separating this pseudo-Debye curve into an anomalous and a normal part, which is again described in terms of just the Debye theory? This argument is perhaps somewhat too sophisticated, but another one must be considered seriously: as yet no explanations have been offered as to the nature of the two kinds of atoms. In the case of the alkali metals it is indeed difficult to imagine to what effect such small energy differences may be due, and the final proof will be given only after this cause has been located. In the case of the more complicated atoms of the carbon group the prospects for locating the origin of the phenomenon are rather better; in the case of diamond an infra-red absorption frequency at $14 \cdot 1\mu$ has been found, whereas the energy difference from the Schottky function (Table XVIII) corresponds to a wave-length of about $13 \cdot 4\mu$, which is in very good agreement. This would mean that the transition from the lower to the higher energy level can in this case be artificially produced by illumination, which should be considered a sound proof for the existence of these levels. Moreover, in the case of Si and Sn the anomalies are well marked, and more recently Ge, in the same group of the periodic system and with the same crystal structure, was found to exhibit a real hump in the specific-heat curve in the temperature region between those of the anomalies in Si and Sn. Unfortunately it is not possible to analyse the anomaly of Ge completely, because it is rather complicated. It appears to be double, consisting of one anomaly of the Schottky type and another more concentrated one (see fig. 13). From fig. 13 it appears evident that this anomaly at any rate cannot be disputed.

There are two more cases known of clearly recognisable and moreover theoretically perfectly explained anomalies of this type. They have also been discovered by Simon and his collaborators. One is solid orthohydrogen, the other gadolinium sulphate; their specific heats are represented in figs. 14 and 15. Although the maximum in both cases lies beyond the lowest temperature of measurement, one has good reason to suppose that the curves will continue in the way described by Schottky. The complicated internal structure of these molecules leads us to expect such effects. Both these examples are of

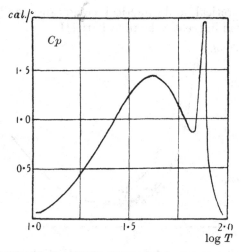

Fig. 13. Specific heat anomaly of germanium.

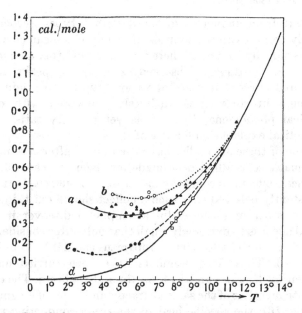

Fig. 14. Specific heat of solid hydrogen (a) 50% Ortho, (b) 75% Ortho, (c) 25% Ortho, (d) 100% Para, $\Theta = 91°$.

great theoretical and practical significance and we shall discuss them in greater detail in Part III.

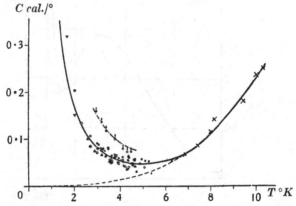

Fig. 15. Heat capacity of gadolinium sulphate. Full line: experimental curve; broken line: extrapolated $T^3$ law; dot and dash line: in a magnetic field of 1800 gauss.

Apart from anomalies of the described types, recognised mainly from a careful analysis of the specific-heat curve by means of Debye's theory, there are numerous other anomalies, which can be directly observed as humps or peaks in the measured curve. They are of various types and origin and, although in some cases definitely connected with other physical phenomena, there is as yet no very satisfactory theoretical explanation for any of them.

Some of these anomalies have been found after careful and systematic search had been made for them. To these belong the heat effects accompanying supra-conductivity. It was persistently believed that such an effect should exist, in spite of a number of unsuccessful attempts to discover it. The actual effect is so extremely small that only after considerable refinement in the calorimetric apparatus could it be found and measured. These measurements have been carried out in Leyden by Keesom and Kok on tin and thallium. The effect for both metals is of the same character and order of magnitude (see fig. 16); the specific heat of the supra-conducting metal follows a smooth curve of not very different character from that for the non-supra-conducting metal, but lying slightly higher;

at the transition point both curves cease abruptly, i.e. the specific heat jumps at this temperature. An extension of the curve for the normal metal can be measured by placing the calorimeter in a magnetic field. Hereby the transition point is lowered and the normal curve can be followed down to the new transition points. Apart from this the measurements on

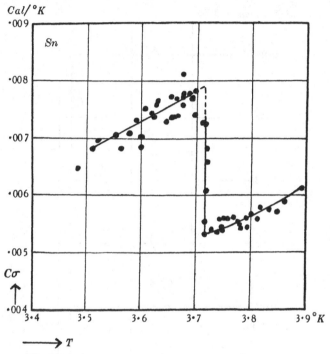

Fig. 16 a. Atomic heat of tin at normal transition point.

thallium have shown a still deeper influence of the field on the character of the heat effect. It becomes more pronounced and takes the form of an almost symmetrical peak, the sharpness and height of which depend on the strength of the magnetic field applied (see fig. 17). This behaviour and the numerical values obtained in these investigations appear to be in good agreement with a theory of supra-conductivity offered by Rutgers, Gorter and Casimir, to which we refer in our general discussion on supra-conductivity in Part IV.

Another group of anomalies which have been looked for in connection with certain physical phenomena are the so-called cryomagnetic anomalies. In the chlorides of chromium, bivalent iron and nickel a temperature dependence of the paramagnetic susceptibility had been observed at normal temperatures, which indicated that at low temperatures these salts

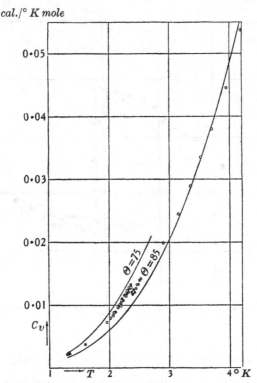

*cal./° K mole*

Fig. 16 *b*. Atomic heat of thallium at normal transition point.

should become ferromagnetic. However, the state in which they are found at these low temperatures is not completely analogous to ferromagnetism. This is sufficient to expect a heat effect analogous to that accompanying an ordinary Curie point of a metal; but as the variation is not uniform for the four salts investigated, it is perhaps not to be expected that the heat effect should be the same for all of them. $FeCl_2$ and $CrCl_3$ have been investigated by Trapeznikova and Schubnikow.

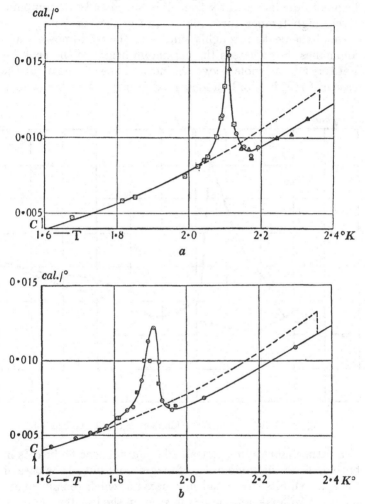

Fig. 17. Atomic heats of thallium in a magnetic field (a) 33·6 gauss, (b) 59·9 gauss.

Their results are represented in fig 18, which shows that in general they correspond to the expectation. The peaks are observed at temperatures somewhat above that at which the extrapolated "Weiss line" intersects the temperature axis, which should be expected from the fact that the susceptibility begins to deviate slightly from this law already at comparatively high temperatures (63° K. in the case of $FeCl_2$). The form of the peaks is roughly similar to the well-known Curie anomalies; but whereas the maximum height of the peak for $FeCl_2$ is 5·6 cal./mole above the normal specific heat, in the case of $CrCl_3$ it is only about 1 cal./mole; the latter peak is

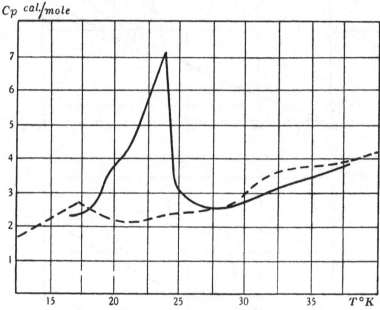

Fig. 18. Molal heat of $FeCl_2$ (full line) and $CrCl_3$ (broken line).

also rather more symmetrical. The normal specific heat is in both cases of the same order of magnitude; but in the case of $CrCl_3$ it exhibits two broad low maxima which remain unexplained. An analogous hump is seen on the low-temperature branch of the peak in $FeCl_2$. Similar phenomena have been observed by the same authors in $CoCl_2$, $NiCl_2$, and $FeBr_2$.

Unexplained humps of various forms and relative magnitudes have been revealed in many investigations. A detailed description is hardly worth while, since in most cases we have not the slightest idea to what cause they may be attributed. It is quite possible that some of these effects, on closer investigation, may lead to the discovery of interesting new phenomena, while others may prove to be a mistake; at the present time they have not yet found special consideration.

This cannot be said of the last group of anomalies, to which we shall turn presently. We refer to the λ-points, which we mentioned already in this Part, chap. I. They have been investigated very thoroughly on various lines and have stood in the centre of discussion for a number of years; yet it must be admitted that we are still far from a satisfactory explanation. For the reasons indicated in chap. I, p. 130, we shall refrain from a discussion of the λ-points in connection with the theory of molecular rotation and confine ourselves to the more thermo-dynamical aspect of the question.

It is an old experience that any physical phenomenon involving heat effects is not clearly understood unless its thermodynamics have first been made clear. In this field lay the difficulties withstanding a successful explanation of the λ-points from the very beginning, and it is but lately that we have begun to see somewhat clearer. The difficulties and the importance of the problem are not well understood unless we are quite familiar with the characteristic properties of the phenomenon. They are in short as follows:

(1) The anomaly extends over a large temperature region; as there is no marked beginning it may be said to extend down to absolute zero.

(2) The anomaly ceases sharply at a temperature which in the most accurately investigated cases could be determined with an accuracy of a few hundredths of a degree.

(3) At this temperature, which we shall call the λ-temperature, the abnormal specific heat has a maximum value, immediately after which it drops to practically zero.

(4) These maxima attain various heights in different instances; they range from a cal./mole to more than *a thousand* cal./mole; in many cases they exceed the normal specific heat several hundred times. Fig. 19 gives a picture of the variety encountered.

⟨ 159 ⟩

(5) The maximum is perfectly sharp; therefore it is difficult to determine its exact height and the values given in fig. 19 must be considered as lower limits.

(6) The accelerated rise of the specific heat on the low-temperature side of the peak is not uniform; in general it is more pronounced for the higher peaks.

A very adequate thermodynamical description of the phenomenon in the neighbourhood of the maximum has been given by Ehrenfest, to whom the invention of the name λ-point is also due. Unfortunately his point of view has frequently been misinterpreted. As we are in the possession of a private letter from Ehrenfest, in which he expresses himself perhaps somewhat more definitely than in his official communication, we feel obliged to his memory to repeat his argument here in full. The following is an almost literal translation of that part of his letter which refers to the λ-points with all emphasis and brackets his.

"On thinking about the *discontinuous* drop in the specific heat of liquid helium at the transition from helium I to helium II I noticed a pretty thing of purely thermodynamical character.

"Singularities of different order in the Zeta-function." $Z(T, P)$ signifies the Zeta-function of a substance, which is considered to be in only *one* phase. I recall the following relations:

1. $-\dfrac{\partial Z}{\partial T} = S,$

2. $\dfrac{\partial Z}{\partial p} = v,$

3. $-\dfrac{\partial^2 Z}{\partial T^2} = \dfrac{C}{T}$ ($C$ = specific heat at constant $p$),

4. $\dfrac{\partial^2 Z}{\partial p dT} = \dfrac{\partial v}{\partial T} = -\dfrac{\partial S}{\partial p},$

5. $\dfrac{\partial^2 Z}{\partial p^2} = \dfrac{\partial v}{\partial p}.$

Now imagine the Zeta-function as a surface above the $p$, $T$ plane. There may be lines along which $Z$ is singular. (For

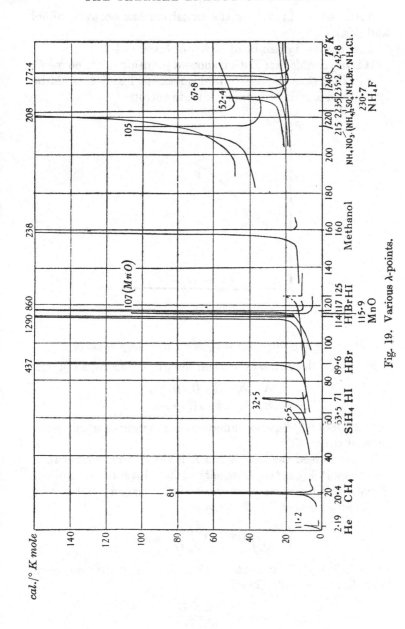

Fig. 19. Various λ-points.

instance when this line is the transition line between liquid and gas.)

The singularity can be of higher or lower order.

It is *impossible* that $Z$ itself shows a jump along the curve $K$. For in that case the differential quotients 1 and 2 would be 'infinite', i.e. infinite volume, infinite entropy.

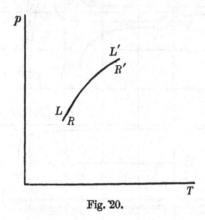

Fig. 20.

*Hence at any rate Zeta itself remains steady at curve K*

However, the Zeta-surface can show a finite kink along $K$:

A. $S_r - S_l = ((S)) \neq 0$,

B. $V_r - V_l = ((v)) \neq 0$.

(The symbol $(( \ ))$ signifies difference at the right- and left-hand sides of curve $K$.)

This is indeed the case, when $K$ is the border line between two phases! From the continuity of Zeta itself the formula of Clapeyron follows at once very nicely: namely $((z)) = 0$ at $L$, $R$ and just so at $L'$, $R'$; therefore

$$\left( \left( \frac{\partial Z}{\partial T} \right) \right) DT + \left( \left( \frac{\partial Z}{\partial p} \right) \right) Dp = 0,$$

where $DT$ and $Dp$ indicate motion *along* the transition curve from $L$, $R$ to $L'$, $R'$. Therefore

$$\frac{Dp}{DT} = \frac{((-DZ/DT))}{((\partial Z/\partial p))},$$

⟨ 162 ⟩

or, on account of 1 and 2:

$$\text{C.} \quad \frac{Dp}{DT} = \frac{((S))}{((v))} = \frac{Q}{T(v_r - v_l)}$$

(formula of Clapeyron with latent heat $Q$). Now singularity of next lower order: not even a kink but only discontinuity in the *second* differential coefficient. (In the curvature.)

Thus because there is *no* kink we now have

$$\text{D.} \quad ((S)) = 0 \quad \text{and} \quad \text{E.} \quad ((v)) = 0.$$

Equation C now leaves $Dp/DT$ perfectly undetermined, because both numerator and denominator become zero.

But it is possible to derive from D and E two equations analogous to C; namely:

$$\text{D'.} \quad \left(\left(\frac{\partial S}{\partial T}\right)\right) DT + \left(\left(\frac{\partial S}{\partial p}\right)\right) Dp = 0,$$

$$\text{E'.} \quad \left(\left(\frac{\partial v}{\partial T}\right)\right) DT + \left(\left(\frac{\partial v}{\partial p}\right)\right) Dp = 0,$$

or

$$\frac{Dp}{DT} = \frac{((-\partial S/\partial T))}{((\partial S/\partial p))} = \frac{-((\partial v/\partial T))}{((\partial v/\partial p))},$$

or the latter equation solved for $\left(\left(\dfrac{\partial S}{\partial T}\right)\right) = \dfrac{C_r - C_l}{T}$ gives

$$\left(\text{for } \frac{\partial S}{\partial p} = -\frac{\partial^2 Z}{\partial T \partial p} = -\partial v/\partial T, \text{ see 4}\right):$$

$$\text{F.} \quad C_r - C_l = -T \frac{((\partial v/\partial T))^2}{((\partial v/\partial p))}.$$

This equation then gives a pretty relation between *discontinuity* of specific heat and discontinuity in coefficient of expansion.

Keesom had derived such a relation by means of cyclic processes, which he showed me. I was interested in finding how all this can be expressed as discontinuity 'of lower order of Zeta'.

What I wanted to understand: These specific heat discontinuities are *similar* to two-phase transitions and yet deeply different from them (no discontinuity in volume, no latent heat!).

Now you see the clear similarity and the clear difference.

In *both* cases 'transition line' with a discontinuity of a differential quotient of the Zeta-function.

*But* in the one instance in the differential coefficient of the *first* order, in the other only in the differential coefficient of the *second order*."

From this quotation it is evident that the criticisms directed against "transitions of the second order" are not to the point, if they aim at proving that transitions between two phases cannot be of the second order. These can indeed not be of any order but the first as long as we deal with a pure substance, i.e. a one-component system. The point of Ehrenfest's argument is just that at a transition of the second order the system does not change phase, but something else, the nature of which is left unexplained. But the thermodynamical behaviour of the system at a second-order transition is well defined and described in all detail by Ehrenfest's equation F. This latter has been very accurately tested for liquid helium by Keesom and for solid methane by Clusius and was found in excellent agreement with experiment (see figs. 20 and 21). This equation F also accounts for the phenomenon observed by Simon and Bergmann in the expansion coefficient of ammonium bromide and phosphate. Instead of a maximum, as in the cases of ammonium chloride, methane and others, the expansion coefficient has a minimum and attains considerable negative values, in other words the density has a maximum at the $\lambda$-temperature and with decreasing temperature the substance expands until a normal contraction appears again, see fig. 22. Now in formula F the expansion coefficient enters in the square, and therefore the effect in the specific heat is independent of the sign of the jump in the expansion coefficient. Liquid helium, by the way, also shows a density maximum at the $\lambda$-point, whereas methane expands abnormally in the same way as ammonium chloride. It is thus very satisfactory that equation F has been tested for substances exhibiting both types of

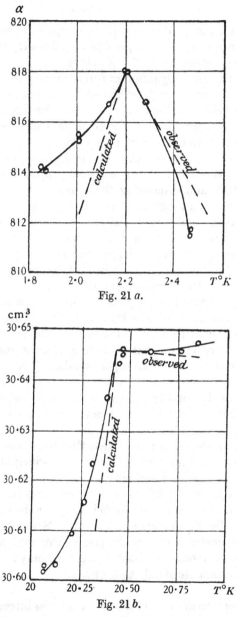

Fig. 21 a.

Fig. 21 b.

Fig. 21. Experimental verification of Ehrenfest's equation. (a) Expansion coefficient of helium, (b) molecular volume of methane.

expansion anomaly. The sum of the experimental evidence considered above appears to provide a sufficiently broad basis for the proposition that the $\lambda$-points are a peculiar new phenomenon principally different from ordinary phase changes as well as from "ordinary" specific-heat anomalies with smooth maxima. We shall for the time being consider them only from this point of view and not concern ourselves with attempts that have been made to explain them as experimental errors. This does not mean, of course, that we rigidly exclude the latter possibility. If experimental material appears of greater accuracy than, and disproving that which we so far possess, this question will undoubtedly have to be raised in full, at any rate in certain instances. However, in the present state of affairs we see no basis for such a discussion. As yet two ways have been recognised by which it should be possible in principle to account for the $\lambda$-anomalies in accordance with thermodynamics, i.e. with Ehrenfest's equation F. One is to assume that the $\lambda$-phenomenon happens in one homogeneous phase, so that the heat effect is to be treated as a true anomaly of the specific heat but with a sharp maximum. The other proposes to treat it as the latent heat of a phase change but with one degree of freedom more than is possessed by a pure substance. Evidently both pictures involve a rather more complex description of the solid body than had hitherto been generally accepted; but it does look at present as though this were unavoidable.

There are in the main two models that have been proposed to account for this complexity; in short, the principal difference between them can be described as follows: either we have to do with a complexity of the molecular constituents of the crystal, such as e.g. assumed by Schottky for the explanation of the anomalies discussed on p. 148, or by Smits in his theory of allotropy, or else the molecules possess this complexity only with reference to the crystal lattice, e.g. they may be associated with a vector capable of different orientations in the framework of the lattice.

The conception of molecular orientation was introduced into the $\lambda$-problem from a rather different domain of physics, the study of solid solutions. To understand this connection we

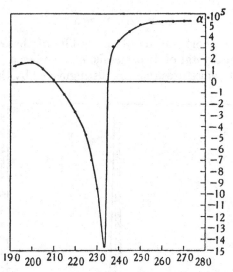

Fig. 22 a. Expansion coefficient of ammonium bromide.

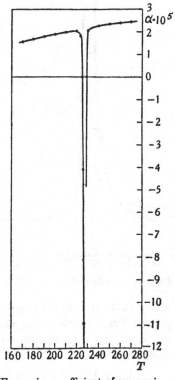

Fig. 22 b. Expansion coefficient of ammonium phosphate.

must give a short outline of the results of these studies. A 50% mixed crystal of two metals, e.g. copper and gold, can exist in different states that are distinguished by the degree of

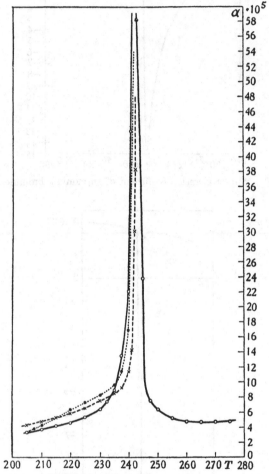

Fig. 22 c. Expansion coefficient of ammonium chloride.

order in which the two sorts of atoms are distributed over the available lattice points. Perfect order means perfect crystallographic regularity, partial disorder ensues when some of the gold atoms exchange their proper places with copper atoms, and perfect disorder is reached when the probability for every atom that its neighbour in a given direction is the one it should be is only one-half. It has been established by X-ray investigations that in a great many similar cases the ordered lattice is stable at low temperatures, strictly speaking at absolute zero, disorder setting in as the temperature is raised. At a temperature which is definitely marked in the state diagram, perfect disorder is attained. Bragg and Williams on the one hand, and Borelius on the other, have developed the theory of this phenomenon on the basis of the assumption that the degree of order is a function of the temperature, and more recently Bethe has computed the extra specific heat to be expected from this theory. He obtains a curve which is very similar to the magnetic specific heat of a ferromagnetic body at its Curie point. Indeed, the conception of Weiss has much in common with that just described; only as a ferromagnetic body is primarily a pure substance, the degree of order is differently determined. In Weiss' picture the magnetic dipoles are, at absolute zero, all pointing in the same direction, whereas above the Curie point they are evenly distributed over all possible directions.

However, the experimental investigations carried out on solid solutions of metals lead to somewhat different results. The two examples that have been studied extensively behave oppositely: the gold-copper alloy undergoes a regular crystal transformation from a tetragonal ordered lattice to a cubic disordered lattice; whereas $\beta$-brass exhibits something much more like a $\lambda$-point than a Curie point. (The maximum specific heat of $\beta$-brass at the temperature where perfect disorder is attained is 300 per cent. of the normal specific heat of the alloy, as against 70 per cent. to be expected from Bethe's theory.)

Now Dehlinger has pointed out that this opposite behaviour, arising evidently from the same cause, can be understood by admitting the possibility of a critical point between two solid phases. The applicability of this conception to the $\lambda$-phenomenon is made evident at the same time. We shall exemplify his idea in a form which may be applicable to the $\lambda$-points.

We assume the molecules in a given crystal to be capable

of two different orientations in the lattice. The lattice is "ordered", if the molecules of different orientation are arranged regularly, so that it can be predicted in what orientation the neighbours of a molecule of given orientation will be. This ordered arrangement can be destroyed with rising temperature, a small effect in the specific heat accompanying the growing disorder. Let the experiment be carried out at constant pressure. It is then possible that at a definite temperature the disorder, although not yet complete, will interfere with the stability of the crystal lattice, and this will undergo an allotropic change into a modification in which complete disorder is at once established. At this temperature a finite latent heat is consumed. The dependence of this temperature on pressure defines the equilibrium curve between the two modifications. Now let us suppose that a critical point exists on this curve, beyond which the two phases merge into one, in analogy to the critical point between a liquid and its vapour. The disorder in this phase will then be established gradually and the lattice will also gradually change its properties; in the specific-heat curve the point where complete disorder is attained is marked by a drop of the heat consumption necessary to bring about the disorder to practically zero. The phenomenon will thus be exactly the same as at a Curie point. At the critical point itself the latent heat disappears; instead we have a true specific heat of infinite magnitude at this point. On moving from the critical point to the Curie points we necessarily pass a region in which the maximum specific heat is extremely high, though not infinite; it grows less the farther we move from the critical point. Thus we can understand the great variety of $\lambda$-points and Curie points that the experiments have displayed (see fig. 19). A detailed quantitative analysis of the specific-heat effects to be expected has been given by Landau. He expands the thermodynamical potential in a series of powers of $\xi$, the degree of order:

$$\phi(\xi) = \phi_0 + \alpha\xi + \beta\xi^2 + \gamma\xi^3 \dots.$$

$\xi$ he defines as

$$\xi = \left(\frac{N_1 - N_2}{N_1 + N_2}\right)^2,$$

where $N_1$ and $N_2$ signify the number of atoms in an ordered and a disordered position respectively. Consideration of the coefficients of the expansion $\phi(\xi)$, which must necessarily be functions of temperature and pressure, leads him to the results described above. Unfortunately he finds it impossible to calculate these coefficients either generally or even in a special case; but he derives a simple formula for the specific heat in the neighbourhood of the $\lambda$-point which can be compared with experiment:

$$C = \frac{\text{const.}}{\sqrt{\tau}}.$$

In this formula the constant is dependent on the coefficients $\alpha$, $\beta$, $\gamma$ and thus on the substance considered; $\tau$ is approximately the temperature distance from the $\lambda$-point. This law appears to hold satisfactorily for the case of $NH_4Cl$.

Of course it should be possible to test experimentally whether there are regions of pressure and temperature in which a substance, ordinarily possessing a $\lambda$-point, undergoes a phase transition.

We notice that these considerations give an interpretation of Ehrenfest's conception of transformations of the second order as changes occurring in a single phase beyond a critical point between two solid modifications. This involves the very important consequence that lattices of different symmetrical properties can be transformed into each other continuously. This seems to be supported by experiment, as we saw in our discussion of crystal structures. However, in all the cases cited there the structural analysis was carried out only at a few temperatures, mostly one below and one above the $\lambda$-point; the conclusion as to the continuity of the two modifications is drawn generally from the specific-heat curve. In many instances we further saw that no change of crystal structure could be detected at all. It thus seems desirable to make a closer study of these substances, including precision measurements with X-rays extending over a large temperature interval.

Only a few such investigations have been carried out, one on sodium nitrate between room temperature and 280° C., and one on MnO between 77° K. and room temperature. Although strictly speaking the former investigation does not come into the theme of this book it is so closely connected with the questions touched upon that we must treat it here.

Sodium nitrate is one of the few substances which possesses a λ-point above room temperature. It crystallises in a layer lattice of the iceland-spar type illustrated in fig. 23. We recognise clearly the layers of Na-atoms interspersed with layers of $NO_3$ groups, all perpendicular to the trigonal c-axis. Fig. 23a comprises only half the elementary cell, the boundaries of which are indicated by double lines. Fig. 23b gives the complete elementary cell, which contains two molecules of $NaNO_3$; it is seen that the $NO_3$ groups are crystallographically different, as they are oriented in different directions. In their discussion of this structure in the "Strukturbericht" Ewald and Herrmann have pointed out that if the O-atoms were to be neglected, the structure could be described with a smaller cell containing only one $NaNO_3$ molecule. We see indeed that if the two $NO_3$ groups in fig. 23b are assumed to be crystallographically equivalent, which is the case if they are rotating about the c-axis or their orientations are randomly distributed within the layers, the spacing along the c-axis is halved. Now Kracek, Posnjak and Hendricks have shown by X-ray investigations that above the λ-point the lattice of $NaNO_3$ conforms to this description, whereas at room temperature the structural difference of the two $NO_3$ groups is quite apparent. They conclude from their Debye diagrams and from their calorimetric experiments, which proved the absence of hysteresis at the λ-temperature, that in the temperature region between 180° and 275° C. the change from the one type to the other takes place continuously. This conclusion was confirmed by dilatometric measurements of Austin and Pierce on single crystals of sodium nitrate. They measured the change in length of a single crystal in different crystallographic directions and found it to take place continuously, the change in the c-axis being abnormally large. As fig. 24 shows, the expansion coefficient in this direction changes with temperature similarly to that of ammonium chloride. These are the only measurements on the expansion coefficient in the neighbour-

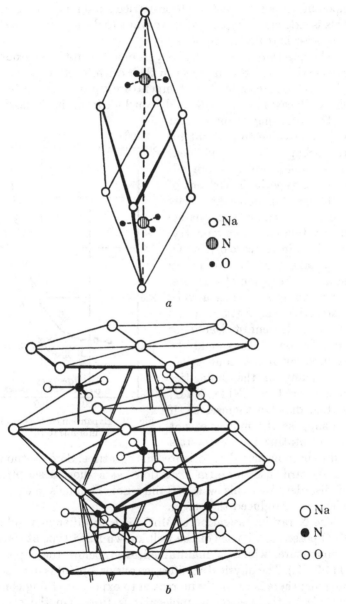

O Na
N
• O

*a*

O Na
• N
o O

*b*

Fig. 23. Sodium nitrate (*a*) elementary cell, (*b*) lattice layers.

hood of a λ-point that have been carried out on single crystals; this is extremely important if we want to decide whether the λ-process is really continuous.

The question whether the $NO_3$ groups start rotating around the c-axis or whether disorder in their orientation in the layer plane takes place with rising temperature is probably pointless. For both cases we can now safely conclude from the discussion in the foregoing chapters that it is impossible to account for the λ-shape of the specific-heat curve unless we make an additional hypothesis like that of Dehlinger, namely that they are connected with a change in crystallographic symmetry, which as in the case of $NaNO_3$ may take place continuously in a certain region of pressure. This example is thus a very instructive verification of the theory of Ehrenfest, Dehlinger and Landau. It should be pointed out in this connection, that many of the difficulties encountered in Table XVII, p. 133, dissolve themselves, if we suppose the molecules not to be rotating in the crystals but to become disorderly oriented in the lattice. Some authors, e.g. Vegard, use the term "rotation" as a short description of disorder, but this was certainly not Pauling's meaning when he introduced that term.

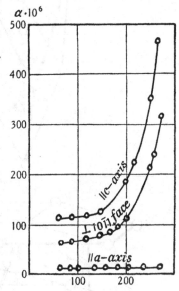

Fig. 24. Expansion coefficient of sodium nitrate.

The X-ray investigation of MnO leads to different results. MnO possesses a lattice of the well-known NaCl type at room temperature, which is maintained even below the λ-point (115·9° K.). The physical significance of this λ-point is not very obvious; there is certainly no reason to expect anything comparable to the λ-effect in molecular lattices. On the other hand, it is also not to be compared with the paramagnetic metals with a Curie point; its paramagnetic susceptibility

follows Curie's law with a negative Curie point. But at the
λ-point the line representing this law suddenly breaks off, and
from about 85° K. on it becomes evident that the susceptibility
now follows Curie's law with a positive Curie point probably
not very far below the temperature to which the measurements extend (see fig. 25).

Debye diagrams were made with a view to determining the
coefficient of expansion. As accurate determinations of the

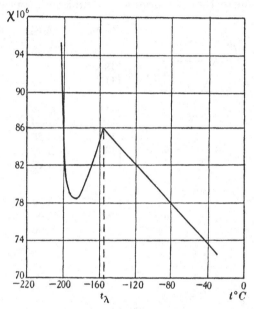

Fig. 25. Magnetic susceptibility of MnO.

lattice constant were needed, lines with very large deflection
angles $\theta$ were registered; the reflection from the crystal plane
(024) with $\theta = 78°$ showed very clearly the resolution of the
$K_\alpha$-doublet of the iron radiation used. The appearance of this
line is the same for all temperatures above the λ-temperature.
Below it undergoes characteristic changes; these are illustrated
in fig. 26, which shows the typical intensity distribution in the
line for four stages that can be roughly discerned. $h_1$ and $h_2$
correspond to the components $\alpha_1$ and $\alpha_2$, $h_3$ to a new maximum
which appears at low temperatures. In the accompanying
table for every film are given: the absolute temperature at

which it was taken, the ratio $h_1/h_2$ and finally the ratio of $h_1$ to the intensity on the same film of the nearest line (004) which shows no resolution of the $\alpha$-doublet.

From this picture, which is drawn according to the results of photoregistration, we conclude the following facts: above the $\lambda$-point the doublet is normal; below the component $\alpha_1$ loses and the component $\alpha_2$ gains in relative intensity. As the spectral distribution in the primary ray remains the same, this can only mean that the original $\alpha$-doublet disappears and a

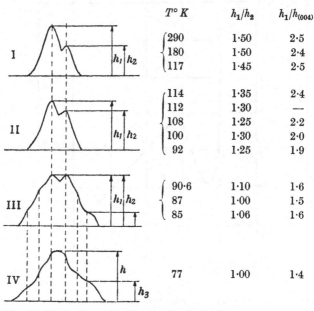

| $T^\circ K$ | $h_1/h_2$ | $h_1/h_{(004)}$ |
|---|---|---|
| 290 | 1·50 | 2·5 |
| 180 | 1·50 | 2·4 |
| 117 | 1·45 | 2·5 |
| 114 | 1·35 | 2·4 |
| 112 | 1·30 | — |
| 108 | 1·25 | 2·2 |
| 100 | 1·30 | 2·0 |
| 92 | 1·25 | 1·9 |
| 90·6 | 1·10 | 1·6 |
| 87 | 1·00 | 1·5 |
| 85 | 1·06 | 1·6 |
| 77 | 1·00 | 1·4 |

Fig. 26. MnO. Intensity distribution in the Debye ring (024) at various temperatures.

new one appears, which is displaced very slightly with respect to the old one, incidentally by an amount which makes the new $\alpha_1$ component coincide with the old $\alpha_2$. The new $\alpha_2$ component is recognised in stages III and IV. This new doublet evidently corresponds to a new lattice constant. The old one was found to be $a = 4\cdot4260$ Å., the new one $a' = 4\cdot4160$ Å. with an accuracy of 0·5 X-units or one-hundredth of a per cent. The difference is relatively considerable: 10 X-units, but still so small that only the lines deflected by large angles $\theta$ show the coexistence of the two lattice constants; the less deflected lines exhibit the change from $a$ to $a'$ only gradually. The important results seen from the line (024) is that both lattice constants appear side by side on the same film under conditions in which thermal equilibrium in the specimen was certainly established. It appears thus that we have here not a gradual transition of the lattice with the constant $a$ into that with the constant $a'$, but that in the temperature interval between 115·9° K. and about 80° K. the two lattices are in equilibrium with each other, the relative quantity of both changing with temperature: above the λ-point the lattice with the constant $a$ is present alone, at temperatures below 80° K. it has practically vanished.

As we saw above the two lattices are characterised besides by their Curie points. The drop in the susceptibility now finds its natural explanation in the fact that in this temperature region we have a mixture of the two lattices.

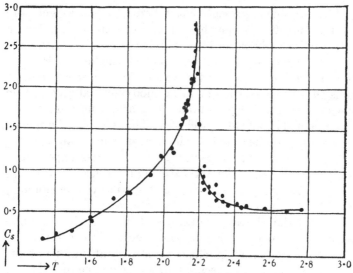

Fig. 27. Specific heat of liquid helium.

Another extremely interesting case which has been studied very thoroughly is the λ-point in liquid helium. All the thermodynamic properties of liquid helium at the $\lambda$-point have been determined with the greatest accuracy by Keesom and his collaborators. Fig. 27 represents the latest results concerning the specific heat. This fits most perfectly to Ehrenfest's description and is the most accurately tested example of a second-order transition. The dependence of the $\lambda$-temperature on pressure was also determined till the intersection of this equilibrium curve with the melting curve. No indication of a critical point was found, but this is only natural as the region in which He II exists is comparatively small, so that the necessary conditions of pressure and temperature cannot be realised in the liquid. The point of intersection with the melting curve is also a singular point of the latter; it marks the deviation of the melting curve in a horizontal direction, owing to which it does not intersect the vapour-pressure curve. Helium is the only substance without a triple point, the only liquid which is stable at absolute zero. The change of the ordinary liquid into that stable down to the zero point of temperature takes place along the $\lambda$-curve.

The $\lambda$-problem presents itself here in a new aspect. In a crystal it is difficult enough to define accurately the difference between the modifications below and above the $\lambda$-temperature; but what are we to propose for a monatomic liquid? How can "order" be interpreted in this case? Nernst's theorem maintains that the transition of a thermodynamical system from one state to another, if it takes place at absolute zero, is not accompanied by an entropy change. Thus if solid helium melts at absolute zero the liquid obtained must be in a state as highly ordered as a crystal. However, Debye-Scherrer exposures of liquid helium show characteristic liquid rings for He I and He II, so that there appears to be no justification in assuming that He II is, in fact, some kind of crystal.

Now all the physical properties of liquid helium, such as density, viscosity, surface tension, dielectric constant, heat conductivity, etc., change characteristically at the $\lambda$-point, but it remains completely obscure how a high degree of order can be maintained in such a liquid. There can certainly be no analogy to the so-called liquid crystals which are common in organic compounds with very large and complicated molecules.

# CHAPTER III

## NERNST'S THIRD LAW

### II. III. 1.  *The Inaccessibility of Absolute Zero*

The fact that the specific heats of solids gradually decrease as the temperature is lowered, until they vanish completely at absolute zero, leads to a most interesting consequence. As a short calculation will make evident, it should enable us to cool a system of solid bodies containing a finite number of moles down to absolute zero.

Consider an adiabatic process, in the course of which the temperature of the system is diminished by $dT$ whenever $dn$ moles react. If $U$ is the molal latent heat of the process and $C$ the total heat capacity of the system, $dT$ and $dn$ are related by the equation*

$$dT = \frac{U}{C}dn. \qquad \qquad \text{......(1)}$$

Now $U$ is equal to the quantity of heat $Q$ which would have to be imparted to the system if the same process were carried out isothermally. If $S$ is the entropy difference and $C'$ the heat-capacity difference of the system before and after the reaction in the isothermal case,

$$Q = TS = T \int_0^T \frac{C'}{T}dT + TS_0,$$

and this integral can be evaluated if we know $C$ as a function of $T$.

Now suppose that we start cooling the system from a temperature so low that $C$ obeys a simple law like that of Debye, e.g.

$$C = aT^b,$$
$$C' = a'T^b,$$
$$b > 1.$$

---

* Here and in future the same thermodynamical symbols will be employed irrespective of the variables. A further specification will be indicated by suffixes, only when necessary.

$S$ then becomes

$$S = \int_0^T \frac{a'T^b}{T}\, dT + S_0 = a''T^b + S_0$$

and

$$Q = T\,(a''T^b + S_0).$$

We may now substitute this value for $U$ in equation (1) and obtain

$$dT = \frac{a''T^b + S_0}{aT^{b-1}}\, dn. \qquad \ldots\ldots(2)$$

Evidently when $T$ approaches zero the numerator tends to $S_0$ and the denominator vanishes. Thus the lower the temperature, the greater will be the thermal effect accompanying the process, and from whatever temperature we start, a finite number of moles will be sufficient to cool the system down to absolute zero by means of any endothermal process.

The existence of a law of nature excluding this possibility was suggested by Nernst in 1905. It is clear from equation (2) that absolute zero would be inaccessible if

$$S_0 = 0.$$

This is in fact the form in which Nernst stated his theorem, thereby relating it to the inaccessibility principle in the same way as the first and second laws of thermodynamics are related to the perpetuum mobile of the first and second kind.

*It is impossible by a thermodynamical process involving a finite number of moles to cool a condensed system to absolute zero.* In other words: In the neighbourhood of absolute zero all thermodynamical processes in condensed systems take place without change of entropy.

Thus in the vicinity of absolute zero the entropy of a condensed system is independent of the physical or chemical state of this system. In order to avoid any possible misunderstanding in future, it will be well to point out that Nernst's theorem does not require all the physical and chemical states in which a system may occur to possess the same free energy at absolute zero: e.g. of two allotropic crystal modifications only one will be "stable"; similarly, a mixture of oxygen and hydrogen crystals will be "unstable" as against a corresponding block of ice, etc. All that the theorem requires is that these states shall be states in the thermodynamical sense of the word, i.e. that entropy and free energy shall be single-valued functions of the independent variables chosen. For such thermodynamically well-defined phases the theorem contains the statement that if between them a reaction, e.g. $H_2 + O = H_2O$, is made to take place at absolute zero, no heat of reaction will ensue, and it is immaterial whether or not this reaction can in point of fact be effected. It is evident that the theorem may not be applied to supercooled liquids, as their free energy is not a univalent function of temperature and volume but depends on time and on the previous treatment of the substance. However, Simon has shown that supercooled liquids are even less effective for producing absolute zero than thermodynamically well-defined substances, since no reversible processes may be performed with them.

As a third law of thermodynamics Nernst's theorem immediately attained immense importance in relation to physical chemistry, in particular to the calculation of chemical equilibria. Hitherto the lack of a guiding principle had made it impossible to calculate the conditions of equilibrium of a given set of substances from their physical properties. Now Nernst's theorem enables us, if the specific heats of all the reagents are known down to absolute zero, to calculate the free energy $A$ of a reaction at all temperatures and thus to determine the equilibrium temperature at which $A = 0$.

Nernst's general formula for $A$ is usually obtained by integrating Helmholtz's equation for the second law:

$$A - U = T \frac{\partial A}{\partial T}. \qquad \ldots\ldots(3)$$

To integrate (3) we divide by $T^2$ and obtain

$$\frac{\partial}{\partial T}\left(\frac{A}{T}\right) = -\frac{U}{T^2}. \qquad \ldots\ldots(4)$$

To calculate the right-hand side we remember that $U$ can be expressed by means of Kirchhoff's law in terms of the molal heats $C$ of the reagents

$$U = \int_0^T \Sigma n C\, dT + U_0, \qquad \ldots\ldots(5)$$

where $n =$ number of moles. The terms under $\Sigma$ are positive for substances formed and negative for those vanishing in the course of the reaction.

Substituting (5) in (4) and integrating we obtain

$$\frac{A}{T} - \left(\frac{A}{T}\right)_{T=0} = -\int_0^T \frac{dT}{T^2} \int_0^T \Sigma n C dT + \frac{U_0}{T} - \left(\frac{U_0}{T}\right)_{T=0}$$

or $\quad \dfrac{A}{T} = -\displaystyle\int_0^T \frac{dT}{T^2} \int_0^T \Sigma n C dT + \frac{U_0}{T} + \left(\frac{A-U}{T}\right)_{T=0} . \quad \ldots\ldots(6)$

$U_0$ can be determined with the help of the specific heats from Kirchhoff's formula (5), if at any temperature $T$ the heat of reaction is known. But the constant $\left(\dfrac{A-U}{T}\right)_{T=0}$ remains undefined, even if we stipulate $A_0 = U_0$, which follows from (3) by merely assuming that $(\partial A/\partial T)_{T=0} = S_0$ remains finite. But Nernst's more rigorous postulate $S_0 = 0$ disposes of this constant and thus relieves us of the impossible task of measuring $A$ down to absolute zero. We may now write

$$A = -T \int_0^T \frac{dT}{T^2} \int_0^T \Sigma n C dT + U_0 . \quad \ldots\ldots(7)$$

The problem of determining $A$ is thus considerably facilitated, provided that Debye's $T^3$ law can be trusted below the temperatures easily accessible to experiment. Deviations from this law at very low temperatures do not in fact preclude the applicability of the third law, but require a more detailed discussion in any special case to prevent us from drawing false conclusions. Some examples will be given later.

The practical importance of equation (7) becomes particularly evident when we try to construct the $A$- and $U$-curves for some reaction as a function of temperature. As we see in fig. 1, after constructing the $U$-curve as indicated above we still have an infinite choice of $A$-curves if we employ Helmholtz's equation, i.e. the second law only. These are the dotted curves which end at absolute zero with finite slope

$$\left(\frac{\partial A}{\partial T}\right)_{T=0} = \left(\frac{A-U}{T}\right)_{T=0},$$

and accordingly with $A_0 = U_0$. From this series Nernst's assumption selects one and only one curve for which

$$(\partial A/\partial T)_{T=0} = 0,$$

and the points of which can be successively constructed from absolute zero up to any temperature by successive application of Helmholtz's equation. The point of intersection of this curve with the $T$-axis, i.e. where $A = 0$, marks the temperature $T_l$ at which the system is in equilibrium.

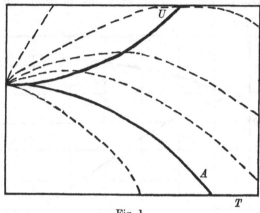

Fig. 1.

We have thus two data, $T_l$ and $Q_l$, the heat of reaction at the equilibrium temperature, which can be very conveniently calculated if they cannot be experimentally determined. For many reactions this is indeed the case; a famous example is the transformation diamond—graphite, where $Q$ can be determined at room temperature by way of the heats of combustion, but $T_l$ is so extremely high that all data referring to this temperature can only be calculated.

On the other hand, simple reactions such as crystal transformations which take place at ordinary temperatures have been successfully used as means of experimental verification of the theorem. The necessary experimental data are available if $T_l$ or $Q_l$ and the specific heats of both modifications can be measured, i.e. if the speed of transformation at the equilibrium temperature is sufficient to make accurate measurements of

$T_i$ and $Q_i$ possible, and at the same time is sufficiently small to allow the high-temperature modification to be supercooled for the measurement of its specific heat.

These requirements are not generally both fulfilled. In some instances the experimental difficulties have, however, been overcome. In fig. 2 we reproduce the results for tin. In this case the equilibrium was obtained by adding a small quantity of grey tin to finely ground white tin powder and keeping this for some weeks at a temperature below 0° C. until all the white tin was transformed into the grey modification. On heating

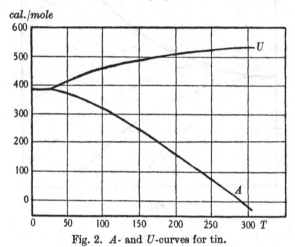

Fig. 2. $A$- and $U$-curves for tin.

slowly the temperature at which the grey tin becomes white again can be fixed within fairly narrow limits. On fig. 2 the $A$-curve calculated from Nernst's theorem as indicated above intersects the $T$-axis within these limits.

We shall not repeat here all the bulk of material that has been brought forward in the course of time to corroborate the theorem. The most convincing proof of the applicability of Nernst's theorem can be seen in the increasing demand on the part of chemical industry of all countries for measurements of specific heats at low temperatures. The material published every day in the scientific and technical journals, especially of the U.S.A., is enormous and even now it could well fill a volume of the size of this one.

This popularity is to be explained by the applicability of the theorem to reactions with gaseous components. These provide the majority of industrial processes, and although the theorem contains an explicit statement on condensed systems only, it nevertheless enables us to draw important conclusions concerning the chemical behaviour of gases.

We recall equation (6), which expresses the second law of thermodynamics in the integrated form

$$\frac{A}{T} = \frac{U_0}{T} - \int_0^T \frac{dT}{T^2} \int_0^T \Sigma n C dT + \left(\frac{A-U}{T}\right)_{T=0}. \quad \dots(6)$$

If this equation is applied to one mole of a single gas, the symbols have the following meaning:

$A = RT \ln p$ is the free energy of the gas at the temperature $T$.

$U_0$ is the total heat of the gas at absolute zero at pressure $p$.

The sum under the integral sign contains only one term, the molal heat $C_p$ of the gas, which may or may not depend on the temperature.

$\left(\dfrac{A-U}{T}\right)_0 = i$ is the entropy constant of the gas.

To find the significance of the same symbols for a reaction of several gaseous components we have merely to sum up the equations for the individual gases, remembering only that the symbols for substances appearing as a result of the reaction enter into the sum with the opposite sign to those of disappearing substances. Thus we find:

$A = RT \sum_r n_r \ln p_r = RT \ln K_p$, where $K_p$ is the reaction constant.

$U_0 =$ the heat of reaction at absolute zero $= \sum_r n_r U_{0r}$.

$\Sigma n C = \sum_r n_r C_{pr}$.

$\left(\dfrac{A-U}{T}\right)_0 = \sum_r n_r i_r = I$, the chemical constant of the reaction.

The determination of the constant $I$ is the central problem of the chemistry of gases. We shall now show that Nernst's theorem enables us to calculate this constant from the thermal data of the condensed form of the gases in question.

Applying equation (6) to the equilibrium of a gas with its liquid or solid, we find in the same way as above:

$A = RT \ln \pi$, where $\pi$ signifies the vapour pressure at $T$.

$U_0 = \lambda_0$, the heat of evaporation at absolute zero.

$\Sigma nC = C_{p \text{ gas}} - C_{p \text{ cond.}}$

$$\left(\frac{A - U}{T}\right)_0 = i_{\text{gas}} - i_{\text{cond.}}$$

Now according to Nernst's theorem

$$i_{\text{cond.}} = 0, \qquad\qquad \ldots\ldots(8)$$

i.e. the vapour-pressure constant of a substance is identical with the chemical constant $i$ (entropy constant) of this gas and *therefore I can be determined as the sum of the vapour-pressure constants of the components.* To find $I$ for a given reaction we must know the same thermal data of the respective solids that we require for any other application of Nernst's theorem, and in addition one vapour pressure at any given temperature for every substance.

## II. III. 2. *The Principal Significance of Nernst's Theorem*

Nernst's theorem contains a definite statement concerning the properties of matter at very low temperatures. The full significance of this statement did not become apparent until, many years later, quantum theory had progressed so far as to make the theorem appear trivial. When Nernst first put forward his third law, the chief interest centred round the fact that it enabled the behaviour of a chemical system to be calculated with the help of calorimetric data obtained at low temperatures. It thus satisfied the practical demands of chemistry and simultaneously stimulated low-temperature research. However, in the course of this research the twofold nature of the theorem became more and more evident. In the controversies which ensued the opponents of the third law gradually separated into two groups: those who held that the law was true but inapplicable and those who maintained that though applicable it was false.

The controversies arose when several cases were found in which the deviations between the values of chemical constants obtained from chemical equilibria and those found with the help of the theorem from the vapour-pressure equation were slightly greater than could be explained from experimental error though much less than could be of importance to practical application. For the accuracy required for chemical calculations, though considerable, is rather less than that of which the highly developed methods of low-temperature calorimetry are capable. At the present time there is little reason to doubt the applicability of the theorem to practical purposes of chemistry, as we have said above. It is indeed difficult to believe that industrial enterprises would run the risk of the costs of low-temperature research unless from their experience they had gained the conviction that it was going to pay.

It was in fact not this sort of applicability that was contested, but the applicability to problems concerning the state of matter at low temperatures, i.e. to purely physical problems. But in this domain the question must be put in quite a different way. The physicist experimenting at low temperatures is not in particular need of a general principle for the calculation of equilibrium conditions. He does not want to study chemical reactions and most crystal transformations at low temperatures take place almost instantaneously, so that equilibrium conditions can only and very conveniently be studied experimentally. The question of applicability in this sense is in fact the same as that of the correctness of the theorem as a fundamental principle, and if one question is answered in the affirmative the other must be too.

To demonstrate this let us tentatively assume that the theorem is correct. What follows for the behaviour of a thermodynamical system at low temperatures? First of all we must, of course, make sure that we actually have to do with a thermodynamically well-defined system. Let us suppose that we want such a system to undergo a certain process by which it is cooled below the temperature $\tau$ we have used hitherto, and we want information about the actual cooling effect obtainable. Knowledge of the specific heat of the substance below $\tau$ is then indispensable, but $\tau$ is the lowest temperature we have hitherto produced experimentally. Thus the question arises: Can we safely extrapolate the specific-heat curve below $\tau$? To answer this question we shall proceed exactly in the opposite way to that which we followed formerly: we shall use the data obtained from chemical equilibria between the substances in question and find out to what degree of accuracy they coincide with those obtained by means of our extrapolated specific-heat curve and Nernst's theorem. If the result is satisfactory we have obtained the necessary corroboration and can go on with our calculation. If we find a discrepancy exceeding that of the experimental error, we have made an important discovery: a hitherto unknown effect causes the specific heat of our substance to deviate from the expected course! This method is extremely sensitive; for at very low temperatures $\tau$ even a very small deviation in $C$ produces a noticeable effect in $A$ since equation (7) contains under the integral sign the expression $C/T$.

A very illustrative example for the fertility of this attitude is the anomaly of solid hydrogen below $9°$ K. (see p. 153). It was predicted by Simon as early as 1926 on the ground of the very argument which we have here set out at length. In his important paper on the chemical constants of several industrially important gases Eucken through very careful and accurate measurements discovered that for a number of reactions the chemical constants were not equal to the sum of the vapour-pressure constants of the component gases. A thorough analysis of the results revealed that these deviations, small and not quite outside the experimental error for most of the gases considered, exceeded these limits definitely in the case of hydrogen, and Eucken drew the conclusion that Nernst's theorem cannot be applied to the computation of reactions involving hydrogen. Now the vapour-pressure constants in these calculations were obtained on the basis of specific-heat measurements of liquid and solid hydrogen extending down to $10°$ K. and subsequent extrapolation of the specific-heat curve to absolute zero with a Debye function $\Theta_D = 91$. Simon on the other hand questioned this extrapolation, maintaining that as Nernst's theorem ought to be considered as an exact law of nature, the specific-heat curve of solid hydrogen should rather be expected to deviate from the Debye curve below 10° K. by an amount compensating the difference between the two determinations of the chemical constant. This expectation found very significant support when it became known on theoretical grounds that hydrogen is by no means the simple substance it was generally believed to be at the time of Eucken's measurements, but is actually a mixture of an ortho and a para modification. It was Eucken himself who almost simultaneously with Bonhoeffer and Harteck succeeded in proving the complex nature of hydrogen experimentally. Shortly afterwards Simon completed his measurements of the specific heat of hydrogen with varying content of para and ortho between 2° and 10° K. The results, which we have already set out on p. 153, revealed indeed a colossal anomaly due to the complex orthohydrogen molecule, which only begins to be noticeable at 9° K. It is impossible to extrapolate the specific heat of the normal hydrogen mixture from measurements which reach down to 10° K.; this is not even possible from 2° K., as the anomalous specific heat at this temperature still tends to rise with falling temperature and it can only approximately be judged that its maximum possibly lies around 1° K.

These discoveries are perhaps the best proof for the correctness as well as the applicability of the theorem. This proof hardly loses in importance by the circumstance that it was not immediately possible to deduct the correct value for the chemical constant of the normal hydrogen mixture from the calorimetric measurements owing to the fact that these will first have to be extended considerably below 1° K. For the necessary correction could be deduced theoretically with great accuracy from the energy levels of the two forms of the hydrogen molecules as revealed from the spectra of the gas.

For the most important diatomic gases the comparison between calorimetrically and spectroscopically derived entropies has been carried out by Giauque and his collaborators. Their results are represented in Table XIX, where $S$ (cal.) signifies the entropy of the gas at its boiling point derived from the specific heat of its liquid and solid by means of

Nernst's theorem, $S$ (spec.) the entropy, also at the boiling point, obtained from the sum of all the energy states which are excited at this temperature. We see from the table that the entropies computed in these two independent ways agree remarkably well except in the cases of CO and NO. These substances are possibly not in stable equilibrium in the solid state.

Table XIX

| Gas | $S$ (cal.) | $S$ (spec.) | Literature |
|---|---|---|---|
| HCl | $41\cdot3\pm0\cdot1$ | $41\cdot47$ | Giauque and Wiebe, **50**, 101, 1928 |
| HBr | $44\cdot9\pm0\cdot1$ | $44\cdot92$ | Giauque and Wiebe, **50**, 2193, 1928 |
| HJ | $47\cdot8\pm0\cdot1$ | $47\cdot8$ | Giauque and Wiebe, **51**, 1441, 1929 |
| $O_2$ | $40\cdot6\pm0\cdot1$ | $40\cdot68$ | Giauque and Johnston, **51**, 2300, 1929 |
| $N_2$ | $36\cdot4\pm0\cdot1$ | $36\cdot42$ | Giauque and Clayton, **55**, 4875, 1933 |
| CO | $37\cdot2\pm0\cdot1$ | $38\cdot32$ | Clayton and Giauque, **54**, 2610, 1932 |
| NO | $43\cdot0\pm0\cdot1$ | $43\cdot75$ | Johnston and Giauque, **51**, 3194, 1929 |

All quotations are from the *Journal of the American Chemical Society.*

This procedure is very general and applicable to any gas whose spectrum has been completely analysed. We are thus in the position to verify Nernst's theorem by data obtained independently of thermal measurements and this is of the utmost importance. For it must be emphasised that the conviction of the validity of the theorem can no more be gained from thermodynamical considerations alone than in the case of the first two laws of thermodynamics.

We can always argue that the corroborations of the theorem are probably more or less accidental, if we admit possible anomalies in the specific heat below the lowest temperature attained and used for practical measurement. The thing to do is to find out whether or not the theorem is in accord with our general physical conceptions. Considering this, we find that Nernst's proposal was indeed an extraordinarily bold one in 1905. Even the older quantum theory, which was then only in its very beginnings, contained nothing to corroborate the theorem. Quantum mechanics alone provides now sufficient ground for our conviction that Nernst's theorem must be regarded as the third law of thermodynamics.

# Part III

## ORBIT AND SPIN

---

### CHAPTER I

### INTERNAL DEGREES OF FREEDOM

#### III. i. 1. *Degenerate States*

The principal step forward from van der Waals' theory of real gases to Debye and Born's conception of solids is that the latter make use of the quantum theory, which becomes inevitable when particles are brought to such close proximity as is the case in solid bodies. However, in both these theories the particles are still considered as billiard balls just as in Boltzmann's original conception of a perfect gas.

Now we know that modern physics has probed far deeper into the structure of matter than we might infer from these billiard-ball theories. We know that the next step was the atomic model of Rutherford and Bohr, which considers the atom as a positive nucleus surrounded by electron shells. Spectroscopy and radioactivity have enabled us to form a fairly adequate picture of the atom with the help first of quantum theory and finally of quantum mechanics.

We have in fact become accustomed to the idea that atoms and molecules have characteristic properties apart from their ability to bounce about in space. We know that they have internal degrees of freedom, and it is to these that we must turn for an explanation of a behaviour which appears anomalous from the standpoint of theories based on the billiard-ball assumption. With these internal degrees of freedom, atoms and molecules are in general capable of occupying a number of discrete quantised states each associated with a definite energy.

We shall now show how this fact will give us a quantitative explanation for one type of anomalous specific heat.

Consider a system of particles capable of occupying two states differing in energy by $\epsilon$. Then, according to Boltzmann's law, the energy of the system at a given temperature $T$ will be given by

$$E = N \frac{\epsilon_1 e^{-\frac{\epsilon_1}{kT}} + \epsilon_2 e^{-\frac{\epsilon_2}{kT}}}{e^{-\frac{\epsilon_1}{kT}} + e^{-\frac{\epsilon_2}{kT}}}, \qquad \ldots\ldots(1)$$

where $\qquad\qquad \epsilon_2 - \epsilon_1 = \epsilon, \qquad\qquad \ldots\ldots(2)$

and we have assumed for the sake of simplicity that the two states have equal statistical weights.

The term in the specific heat due to the existence of these two states will therefore be

$$\Delta C = dE/dT = \frac{N}{kT^2} \frac{(\epsilon_1 - \epsilon_2)^2 e^{-\frac{\epsilon_1 + \epsilon_2}{kT}}}{(e^{-\frac{\epsilon_1}{kT}} + e^{-\frac{\epsilon_2}{kT}})^2}.$$

Inserting (2) and putting $\frac{\epsilon}{k} = \Theta$ and $N\epsilon = U$, we obtain

$$\Delta C = \frac{R \left(\frac{\Theta}{T}\right)^2 e^{\frac{\Theta}{T}}}{(e^{\frac{\Theta}{T}} + 1)^2}.$$

This formula, which was first deduced by Schottky, leads to a specific-heat curve of the type shown in fig. 10, Part II, chap. II. Its "explanation" is simple: at absolute zero all the particles will be in the lower state; at infinite temperature they will be evenly distributed over both. At intermediate temperatures an input of heat will serve to effect a certain number of transitions from the lower to the higher state, and we might deduce that the largest number of transitions will occur around $T = 0.4\,U/R$. In this region we may therefore expect an anomaly in the specific-heat curve, as the heat must lift particles into the higher state as well as raise the temperature. Moreover, it is clear that at temperatures far below and far above $U/R$, thermal measurements will fail to reveal the existence of several states. Only at and around the characteristic temperature will the true state of affairs become manifest. It is easy to generalise this concept for cases when more than two states are possible or when the various states have different statistical weights. An example will be given in the next paragraph.

Now the transitions associated with what is usually termed the coarse structure of spectra and which give rise to the emission or absorption of visible and ultra-violet light possess characteristic temperatures that can hardly be produced in the laboratory; e.g. the frequency corresponding to a wavelength of 5000 Å. is $0.6.10^{15}$ sec.$^{-1}$. Multiplying by Planck's constant we have $h\nu = 3.6.10^{-12}$ erg/atom. Putting $h\nu = k\Theta$, where $\Theta$ is the characteristic temperature, and converting the units into calories per gramme-molecule, we obtain

$$U = R\Theta = 0.5.10^5 \text{ cal./mol.}$$

Thus $\qquad\qquad \Theta = 2.10^4 \,^{\circ}\text{K.}$

The thermal excitation of these processes is therefore impossible even at the highest available temperatures. If, however, we turn to the fine and hyperfine structure of spectra, we come to very different results. More especially the hyperfine structure is caused by transitions of so small an energy output that its characteristic temperatures lie somewhere around a thousandth of a degree. Thus at all temperatures hitherto attained, apart from those we shall speak of at the end of this Part, the atoms are already evenly distributed over the respective states.

If a particle or system of particles is capable of occupying two or more states possessing the same energy, these states are known as degenerate. There are a number of ways of expressing this fact. According to statistical mechanics a state is degenerate if more than one configuration exists which leads to it. In wave mechanics a state is defined as degenerate if its eigenvalue satisfies more than one eigenfunction. Theorists have shown that all these definitions are synonymous if taken in the right spirit.

Now if we suppose an external field to act on a system of particles occupying a degenerate state, the energy of each particle will as a rule be changed. Moreover, the influence of the field on the several states that are degenerate with respect to one another will in general differ. Thus, under the influence of an external field, states which were originally degenerate may lose this property. The term corresponding to the degenerate state will "split up". Typical examples of this occurrence are the Zeeman and Stark effects. However, it is not always an external field which splits up a degenerate state. In many cases the field engendered by neighbouring molecules will suffice, and we then speak of degeneracy as removed by interaction. The degree to which the degenerate state is split up naturally depends on the magnitude of the field. The weaker the field, be it external or the result of interaction, the nearer together lie the energies of the erstwhile degenerate term. It is an open question whether all degenerate terms are in reality split up by some field or other. It is natural to suppose that a certain amount of interaction always exists and that if the state still appears to be degenerate it is because the splitting up is too small to be detected. It has become customary to call states degenerate when at ordinary temperatures and in the absence of an external field there is no way of detecting that they are not. When at low temperatures the energy difference becomes apparent, we say that at these temperatures the degeneracy is removed. This is only another way of saying that in reality the states were never degenerate at all.

Nernst's theorem, which in its original thermodynamic form postulated that entropy differences should vanish at absolute zero, must from the point of view of statistical theory be interpreted as signifying that the ground states of atoms and molecules cannot be truly degenerate. For degeneracy of the ground level means that this state can be realised in more than one way, and thus has a thermodynamical probability greater than unity, and this would lead to a finite zero-point entropy.

### III. i. 2. *Ortho and Para Hydrogen*

To determine how the various energy states of atoms and molecules contribute towards the specific heats and more generally towards the thermodynamical functions of a system, we may employ the well-known statistical equations connecting these functions with the energies of the states. Thus the entropy contribution $S$ of a number of states with energies $\epsilon_\nu$ is given by

$$S = R\,(\ln Q + T\,.d\ln Q/dT),$$

where

$$Q = \Sigma p_\nu e^{-\frac{\epsilon_\nu}{kT}}.$$

From what we have said concerning degenerate states the statistical weights $p_\nu$ signify groups of states, the energies of which are very similar and which it is therefore profitable at the temperature concerned to consider as one $p$-fold degenerate state.

The specific heat $C$ can be computed from this formula with the help of the relation

$$C = T\,.dS/dT.$$

These formulae are in principle not restricted to perfect gases; however, it is only for these that the values of the $\epsilon_\nu$, needed to compute the thermodynamical properties, can as a rule be determined. Let us take as an example the case of diatomic gases. The discrete energy states of a diatomic molecule are connected with their several degrees of freedom. These are nuclear, rotational, vibrational and electronic in the order of increasing energy. The energy levels of the various states have been determined with great accuracy with the methods of spectroscopy. Using the above equations Giauque and his collaborators were thus able to calculate the thermodynamical functions of gases, more especially the entropies, free energies and specific heats, in so far as they are due to the excitation of these energy levels. Superposing their results upon the terms of the functions dependent on translational motion, they could thus determine the total entropy.

Now the energies needed to excite vibrational, and still more electronic levels, are in almost all cases so high that at ordinary temperatures their excitation may be neglected. Moreover, with a single exception, the rotational energies are so small that at the lowest temperatures at which the gas can as such exist the molecules have long since reached equipartition with respect to them, and at still lower temperatures at which the substance is liquid or solid, the states are not known. The one exception to this rule is hydrogen, which for this and other reasons has given rise to endless discussions and much experimental research.

As a result of the small moment of inertia of the hydrogen molecule and the low boiling point of the gas, the excitation of the rotational energy levels occurs at temperatures at which hydrogen is still in the gaseous state. At the boiling point practically all the molecules are in the lowest rotational state and equipartition sets in shortly below room temperature. We can therefore compute the rotational specific heat of hydrogen and test the result by direct experiment. Now according to Eucken's very accurate measurements a marked discrepancy exists between the experimental values and those obtained from the above equation. This discrepancy could be explained only with the concepts of quantum mechanics, taking into account the spin of the nucleus.

In sketching the line of thought which has led to a solution of the problem it will be well to generalise sufficiently so as to include the case of deuterium as well as ordinary hydrogen. Consider a diatomic molecule with two identical atoms. Then the eigenfunction $\psi$ describing the molecule can be expressed as a product of three eigenfunctions $\phi_e$, $\phi_r$ and $\phi_n$, referring respectively to the electron motion, the rotation of the molecule and the spin of the nuclei. According to the generalised Pauli principle the $\psi$-functions corresponding to the various states of the molecule must be either all symmetrical or all antisymmetrical with respect to the two nuclei. Which group of states actually occurs in nature cannot be determined a priori, but we know from experience that in some particles one group occurs and in some the other. In the case of the two hydrogens we can be certain that the ground electronic state possesses an eigenfunction $\phi_e$ which is symmetrical in the nuclei; and at all temperatures obtainable in the laboratory only this state is occupied. In considering the symmetry properties of the molecule, those due to $\phi_e$ therefore cancel out, and we may thus confine ourselves to a discussion of the other two.

Let us begin with the function $\phi_n$ which characterises nuclear spin. If the value of the nuclear spin of an atom be designated as $t$, the quantum number $S$, which characterises the total nuclear spin of the molecule, may assume the $2t + 1$ values $2t$, $2t - 1$, $2t - 2$, ... 0. Moreover, for every value of $S$ there are $2S + 1$ possible eigenfunctions. Now according to quantum mechanics all the eigenfunctions associated with quantum numbers $S = 2t - n$ with odd values of $n$ are anti-symmetrical with respect to the nuclei, whereas the functions corresponding to even $n$ are symmetrical. Terms which are antisymmetrical with respect to the nuclei are known as *para* terms, the symmetrical terms being called *ortho*. The selection rules of quantum mechanics show that the pro-bability of transitions from ortho to para terms and vice versa is very much smaller than that for transitions from ortho to ortho and from para to para. A substance capable of assuming ortho and para states must therefore behave to all intents and purposes like a mixture of two separate substances.

At sufficiently high temperatures, at which the molecules may be considered as evenly distributed over all possible states, the relative concentration of ortho to para molecules is given by the quotient of the total number of ortho eigen-functions over the total number of para eigenfunctions, i.e.

$$c_0/c_p = \frac{\sum_{n\,\text{even}} 2(2t - n) + 1}{\sum_{n\,\text{odd}} 2(2t - n) + 1} = \frac{(t+1)(2t+1)}{t(2t+1)} = \frac{t+1}{t}.$$

......(1)

Turning to the rotational eigenfunction $\phi_r$, we know that all $\phi_r$ associated with an even value of the rotational quantum number $j$ are symmetrical in the nuclei and all $\phi_r$ corresponding to odd $j$ antisymmetrical. Hence we may deduce the following schedule:

| Antisymmetrical | | Symmetrical | |
|---|---|---|---|
| $\phi_r$ | $\phi_n$ | $\phi_r$ | $\phi_n$ |
| $j = 0, 2, 4, ...$<br>$j = 1, 3, 5, ...$ | Para<br>Ortho | $j = 0, 2, 4, ...$<br>$j = 1, 3, 5, ...$ | Ortho<br>Para |

I.e., if the total function is anti-symmetrical para terms are associated with even rotational quantum numbers and ortho terms with odd ones; if $\psi$ is symmetrical it is the other way about. Moreover, in the first case the lowest rotational term with $j = 0$ is a para state, whereas in the second case it is an ortho state. Thus in the first case pure para will be in equilibrium at the lowest temperatures, in the second case pure ortho. However, as transitions between para and ortho terms are assumed to be very rare, we may suppose that at low temperatures equilibrium is in general not reached and that in both cases para and ortho molecules may exist side by side at all temperatures.

If we know the value of $t$ we can determine the weights and thus the entropies, specific heats, etc. for various cases. This was done for ordinary hydrogen by Beutler, taking $t = \frac{1}{2}$, as known from Stern's experiments with molecular rays.

Beutler computed the specific heat of pure ortho and pure para hydrogen, assuming that in the case of hydrogen the total eigenfunction is antisymmetrical. Apart from these he calculated the specific heat of an equilibrium mixture, the concentration varying with the temperature, and of a mixture of the concentration 25 per cent. para and 75 per cent. ortho (which, from equation (1), should be in equilibrium at high temperatures), assuming that no para-ortho transitions occur. The last curve was found to be in excellent agreement with Eucken's measurements, thus showing that no ortho-para transitions take place within the period of an experiment.

Shortly afterwards the para-ortho theory was finally proved by independent experiments of Bonhoeffer and Harteck and Eucken and Clusius. Whereas the latter showed that by keeping hydrogen at low temperatures for some weeks the specific heat changed in accordance with theory, the percentage of parahydrogen gradually increasing, Bonhoeffer and Harteck succeeded in catalysing the ortho-para transition on charcoal at low temperatures, thus markedly accelerating the reaction. By determining the thermal conductivity, which is closely connected with the specific heat by a simple formula, they demonstrated that almost pure parahydrogen can be obtained in a few minutes by sorbing on charcoal at 20° K. Indeed, Beutler's calculations show that at the normal boiling point of hydrogen only about $\frac{1}{10}$ per cent. of ortho is in equilibrium with the para component.

Recently a similar set of experiments was carried out on deuterium by Clusius and Bartholomé and by Farkas and Farkas. The results show definitely that the nuclear spin of deuterium is equal to 1, and that here the total eigenfunction is *symmetrical* in the nuclei. At high temperatures we thus have 66·6 per cent. ortho in equilibrium with 33·3 per cent. para, whereas at low temperatures pure orthodeuterium prevails.

We have seen that for every value of the nuclear quantum number $S$ there are $2S + 1$ eigenfunctions. A state corresponding to $S$ is therefore $2S + 1$-fold degenerate. Moreover, a rotational state associated with a rotational quantum number $j$ possesses a degree of degeneracy amounting to $2j + 1$. We thus obtain the following table for the weights of the various rotational terms of ordinary hydrogen:

| $j$ | $=$ | 0 | 1 | 2 | 3 | ... |
|---|---|---|---|---|---|---|
| Term | $=$ | $p$ | $o$ | $p$ | $o$ | ... |
| $2S+1$ | $=$ | 1 | 3 | 1 | 3 | ... |
| $2j+1$ | $=$ | 1 | 3 | 5 | 7 | ... |
| $g$ | $=$ | 1 | 9 | 5 | 21 | ... |

Whereas the lowest state of parahydrogen is undegenerate, the next lowest ortho term is ninefold degenerate, threefold with respect to rotation and threefold with respect to nuclear spin. Moreover, we know that orthohydrogen may be arbitrarily supercooled. If this degeneracy were to subsist down to absolute zero, orthohydrogen should possess a finite zero-point entropy and Nernst's theorem would be violated. The specific heat of solid hydrogen, which was demonstrated in fig. 14 of Part II, chap. II, shows that we have no reason to doubt the validity of Nernst's theorem. We saw that the specific heat of parahydrogen remained normal to the lowest temperatures attained, whereas that of orthohydrogen showed a marked anomaly below 9° K., the maximum of which could not yet be determined. We must therefore assume that in solid orthohydrogen the energy differences between the components of this quasi-degenerate term are such that they become noticeable below 9° K. Though the maximum of the anomaly could not yet be fixed, we may conclude by a rough extrapolation that it will be at temperatures of the order of a degree. From theoretical considerations we may infer that this will be a likely characteristic temperature for transitions between the terms of the rotational triplet, whereas the spin triplet will probably split up at much lower temperatures. How this and similar expectations are interwoven with the problem of attaining still lower temperatures will be discussed in chap. III of this Part.

# CHAPTER II

# PARAMAGNETISM

### III. II. 1. *Degeneracy, Magnetic Moment and Zeeman Effect*

The existence of degenerate or rather quasi-degenerate states is of far more general importance than our example of para- and orthohydrogen might lead us to suppose. It is indeed at the bottom of all paramagnetic phenomena. Paramagnetism is not only one of the chief fields of research in low-temperature physics but has also shown us the most potent instrument for producing very low temperatures.

Consider an atom in a certain quantum state $n$ characterised by the energy $W_n^0$. If the atom be placed in a magnetic field the energy *may* be changed, and can then be written in the form of a series in $H$, which we shall break off after the first two terms:

$$W_n^H = W_n^0 - M_n H + A_n H^2.$$

As the magnetic moment is defined by $-dW/dH$, we see that the latter consists of a term $M_n$, which is independent of the field strength, and an "induction" term $-2A_n H$. We shall here consider only the first of these terms and confine ourselves to small values of $H$.

In the general case, when the atom is capable of assuming a large number of states $n$ with different magnetic moments, the effective magnetic moment per atom of a gas consisting of such atoms is found according to the well-known laws of statistical mechanics by averaging over all the states in the same way as was done in the last chapter. We thus obtain

$$\bar{M}_H = \frac{\sum_n M_n e^{-\frac{W_n^H}{kT}}}{\sum_n e^{-\frac{W_n^H}{kT}}}. \qquad \ldots\ldots(1)$$

Now if we assume that the total moment of the gas in the field zero vanishes, the above expression may be transformed and written in the following form:

$$\bar{M}_H = \frac{H}{kT} \frac{\sum_n M_n^2 e^{-\frac{W_n^0}{kT}}}{\sum_n e^{-\frac{W_n^0}{kT}}}. \qquad \ldots\ldots(2)$$

The magnetic moment of the gas is thus reduced to an expression depending on the field strength, on the magnetic moments of the atoms in the various energy states and on the energies of these several states in the absence of a magnetic field. For simplicity we shall begin with the case that all excited states of the atom are very high compared with $kT$. We might now suppose that the sums would reduce to a single term and we could write

$$\bar{M}_H = H M_n^2 / kT.$$

This would lead to definitely false results. To understand why, we must refer to the concept of degenerate states we discussed in the last chapter.

The three quantum numbers that characterise the magnetic behaviour of the atom are the orbital quantum number $L$, the spin quantum number $S$ and the quantum number $J$, which defines the total moment of momentum and is a combination of $L$ and $S$. Now just as in the case of the hydrogen molecule when $j$ characterised the total moment of momentum of the molecule, a state denoted by $J$ is $2J + 1$-fold degenerate. It is the splitting up of such terms into $2J + 1$ undegenerate components in a magnetic field that gives rise to the Zeeman effect. The magnetic moment of the atom in a magnetic field is then determined by a magnetic quantum number $M$ which can assume any of the $2J + 1$ values $J$, $J - 1$, $J - 2$, ..., $-J$. We therefore see that even when our atom is capable of assuming only one state with a definite value of $L$, $S$ and $J$, we have still to extend the summation in (2) over the $2J + 1$ Zeeman components into which the term has been split up by the field.

PARAMAGNETISM

Now the magnetic moment $M_n$ is connected with the quantum numbers $L$, $S$, $J$, $M$ by the relation

$$M_n = \mu_0 g M, \qquad \ldots\ldots(3)$$

where
$$\mu_0 = \frac{eh}{4\pi mc}, \qquad \ldots\ldots(4)$$

and $g$ is Landé's factor

$$g = \frac{3}{2} + \frac{S(S+1) - L(L+1)}{2J(J+1)}. \qquad \ldots\ldots(5)$$

$\mu_0$ is known as Bohr's magneton, and is the magnetic moment an atom with one outer electron would have if the latter possessed no spin.

Thus, for our special case of one $2J+1$-fold degenerate quantum state, equation (2) reduces to

$$\bar{M}_H = \frac{\mu_0^2 g^2 H}{kT} \sum_{M=-J}^{+J} M^2 = \frac{J(J+1)g^2\mu_0^2 H}{3kT}. \qquad \ldots\ldots(6)$$

The susceptibility is defined by the magnetic moment of a given quantity of a substance divided by the field producing this moment and may thus be written $\chi = M_H/H$.

In our case
$$\chi = \frac{J(J+1)g^2\mu_0^2}{3kT}. \qquad \ldots\ldots(7)$$

For the special case considered we have thus deduced Curie's law: The susceptibility of a paramagnetic gas is inversely proportional to the temperature. The constant $C$ in the equation
$$\chi = C/T,$$

which is called Curie's constant, may be written

$$C = \frac{J(J+1)g^2\mu_0^2}{3k}. \qquad \ldots\ldots(8)$$

When more than one state must be taken into account, the summation in equation (2) is more complicated. We shall not continue the analysis, which would in any case be fragmentary. In point of fact there are no paramagnetic monatomic gases. The paramagnetic gases that we know are all diatomic and most paramagnetic bodies are not gases at all but solids, for which the theory is exceedingly complicated. At this point it will be well to emphasise the main fact which equation (6) conveys and which is retained in all the other less transparent cases. Paramagnetic susceptibility and a linear Zeeman effect are synonymous; both are due to the occupation of a $2J+1$-fold degenerate state with $J > 0$.

In the case of a molecule consisting of two atoms the quantum number $L$ must be replaced by its projection $\lambda$ along the axis of the molecule. The terms of the molecule, and thus also its behaviour in a magnetic field, are dependent on the coupling forces between orbit and spin. According as $\lambda = 0$, 1, 2, ... the term is called $\Sigma$, $\Pi$, $\Delta$, etc. In every case the total moment of momentum $J$ is formed by combining the moments of momentum of the orbit, the spin and the rotation of the nuclei. If the coupling forces between orbit and spin are great, $J$ is the resultant of the nuclear momentum $V$, the projection $\lambda$ of $L$ along the axis and the projection $\sigma$ of $S$ along the same axis (case $a$). If the coupling forces are small, $J$ is formed by vectorial addition of $V$, $\lambda$ and $S$ (case $b$). This case is always realised in $\Sigma$-states, i.e. when $\lambda = 0$.

The magnetic susceptibility of a gas consisting of diatomic molecules can be computed for the following cases:

(1) Ground level of the type $^1\Sigma$, i.e. $\lambda = \sigma = 0$. The gas is diamagnetic.

(2) Ground level a $\Sigma$-term but $\sigma \neq 0$. Case $b$ is valid. The term is a very narrow multiplet and the magnetic moment is due to the spin only:

$$\chi = \frac{4\mu_0^2 S (S+1)}{3kT}. \qquad\qquad \ldots\ldots(9)$$

(3) $\lambda \neq 0$. Case $b$. $\quad \chi = \dfrac{\mu_0^2}{3kT}(4S(S+1)+\lambda^2).$ $\qquad \ldots\ldots(10)$

(4) $\lambda \neq 0$. Case $a$. If the ground multiplet is narrow in comparison with $kT$ we have formula (9) again. If the ground multiplet is wide as compared with $kT$,

$$\chi = \frac{\mu_0^2}{3kT}(2\sigma+\lambda)^2. \qquad\qquad \ldots\ldots(11)$$

## III. II. 2. *Magnetic Measurements at Low Temperatures*

All methods for measuring magnetic moment or susceptibility are based on the fact that in a non-homogeneous field a
paramagnetic body is acted upon by a force in the direction
of the greatest field strength. This force is proportional to the
non-homogeneity of the field and the magnetic moment of the
body, which is itself, as we have seen, generally proportional
to the field strength. If $\rho$ is the density of the substance, $q$ the
cross-section of the body perpendicular to the field and $ds$ a
differential distance in the direction of the field, a quantity of
substance $\rho q \, ds$ having a magnetic moment $\mu\,(H)$ per gramme
will in a field $H$ incur the force

$$F = \rho q \, ds \, \mu\,(H) \frac{dH}{ds}. \qquad \ldots\ldots(1)$$

If $\mu$ is everywhere proportional to $H$ this formula may be
written

$$F = \rho q \, ds \, \chi H \frac{dH}{ds}. \qquad \ldots\ldots(2)$$

In general the substance to be investigated is suspended
between the poles of an electromagnet of known field strength
and non-homogeneity and the force measured by means of
weights or electromagnetic compensation.

If the sample to be investigated is *very small* the field strength
at every point may be considered as the same. In this case
$\rho q \, ds$ in (1) may be replaced by the mass $m$ of the sample and
(1) and (2) may be respectively written

$$F = m\mu\,(H) \frac{dH}{ds}, \qquad \ldots\ldots(3)$$

$$F = m\chi H \frac{dH}{ds}. \qquad \ldots\ldots(4)$$

The sample should then be placed at the point where
$H\,(dH/ds)$ is a maximum. If larger quantities of the substance
are to be employed, it is profitable to use a sample in the form
of a rod, one end of which is placed in a position where the
field strength is a maximum and homogeneous, the other end

being at a point where $H$ is practically zero. Formula (1) then becomes

$$F = \rho q \int_0^l \mu(H) \frac{dH}{ds} ds = \frac{m}{l} \int_{H_0}^{H} \mu(H) dH \quad \ldots\ldots(5)$$

and formula (2)

$$F = \int_0^l \rho q \chi H \frac{dH}{ds} ds = \frac{m\chi}{2l}(H_l^2 - H_0^2). \quad \ldots\ldots(6)$$

Here $l$ signifies the length of the sample, $H_l$ the field strength in the homogeneous region and $H_0$ the field strength at the other end of the sample.

With the latter method it is usually possible to obtain larger forces. However, it has disadvantages when $\mu$ is not proportional to $H$. For whereas according to formula (3) $\mu$ as a function of $H$ may be determined by measuring $F$ at different values of $H$, this is not possible according to formula (5).

At room temperatures the forces that can be produced are comparatively small, and at the best of the order of magnitude of a few grammes weight. At low temperatures, however, with increasing susceptibility, the forces become greater, and at very low temperatures so considerable that they are now leading to the development of a new technique. In the experiments that concern us at this point the accurate measurement of susceptibilities requires large magnetic fields and thus necessitates the use of large and expensive electromagnets. The largest Leiden magnet produces a field of 30,000 gauss across a pole gap of 16 mm. and requires 450 kilowatts. However, field strengths of such magnitude are not always needed and much valuable work has been done with smaller magnets. Indeed, in some cases small fields are essential, as we shall see presently.

The necessity of bringing the poles close to the sample in order to produce high field strengths is one of the chief problems of low-temperature magnetic work. For in order that the substance may be cooled to the desired temperature a Dewar vessel is indispensable, and in some cases, when very low temperatures are required, several Dewar vessels, one inside the other are needed. Each Dewar vessel necessitates four walls between the poles, not to mention the spaces between the walls and between the innermost wall and the substance itself, which, as we shall presently see, must be suspended freely without touching the sides of the vessel. On the other hand, a sufficient store of cooling liquid must be contained in the Dewars, so that these must have a considerable capacity. As a result a characteristic type of Dewar vessel has been devised for magnetic measurements, which is wide above and very narrow below. The lower part is placed between the poles of the magnet, while the upper part, situated above the pole shoes, contains a reserve storage of cooling liquid. Great ingenuity has been shown on the part of glass-blowers, especially at Leiden, in devising and making the narrow parts of these Dewars. Double and even treble Dewars have been blown in one piece, the entire width being less than 2 cm. Naturally the silvering and evacuation of these vessels, the walls of which are very thin and very close together, demand a considerable technique.

In the experiments carried out by Weiss and his collaborators and by the Leiden school, much labour has been devoted to devising apparatus suitable for work under various conditions. According as gases, liquids or solid bodies were to be investigated, substances with high or low susceptibility, substances that could be obtained in large samples or of which only a few milligrammes were procurable, in cases when a linear or a more complicated relation between magnetisation and field strength was to be expected, different experimental methods were employed. These methods differ mainly in the device for determining $F$.

(1) In the "classical" form of apparatus used by Weiss at room temperature the force is determined by suspending the sample on a fine thread from one arm of an accurate balance, the pan of which has been previously removed. The sample is counterpoised by weights placed on the other pan before applying the magnetic field. The additional force exerted by the field is then compensated by adding the necessary weights.

This method is not easily adaptable to low-temperature work, since the vessel in which the sample must be suspended within a cryostat cannot be sealed off from the balance. Thus the latter must either be in a vacuum or else contain a constant atmosphere of some diamagnetic gas. Though the manipulation of weights *in vacuo* presents no very serious technical difficulties, the apparatus is usually too sensitive to enable this to be done without destroying the adjustment. For this reason it has been found preferable to compensate the magnetic forces by electrodynamic attraction or repulsion. This device has lately been employed very successfully by Wiersma, Woltjer and de Haas.

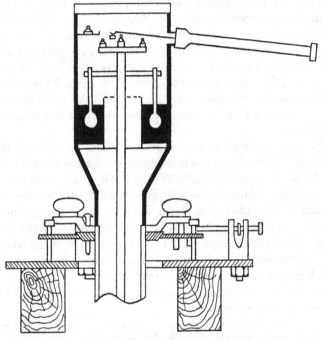

Fig. 1. Magnetic balance on mercury floats.

(2) In a form of apparatus frequently used at Leiden (see fig. 1) the specimen is suspended from a system resting on several glass floats immersed in mercury. The magnetic force produces a slight lift, which can be measured, and which is proportional to the force.

(3) To determine the susceptibility of gases at low pressures, the volume susceptibility of which is naturally very small, de Haas, Wiersma and Capel employed a very useful and attractive method. A small hollow sphere of thin glass was

suspended in an atmosphere of the gas to be studied within a Dewar vessel containing the cooling liquid. The narrower lower part of the Dewar vessel was placed in the magnetic field so that the glass bulb was situated where $H\,(dH/ds)$ was a maximum in the horizontal plane of symmetry of the field (see fig. 2). The susceptibility of the bulb itself was very small, so that when the field was applied the surrounding paramagnetic gas, attracted towards the maximum of $H$, caused the bulb to be displaced away from the field in a horizontal direction, until the displacement was compensated by gravity. By a special device the bulb could then be brought back to its original position and the displacement registered.

(4) The principle of letting the magnetic force cause a displacement in a horizontal direction has great advantages, as in this way comparatively large displacements may be obtained with very small forces. The same principle was frequently employed by Weiss in experiments at room temperature. Woltjer and Wiersma adapted the method to low-temperature work on compressed gases. In these experiments the gas was contained in a thick-walled copper tube, the "large sample method" (see p. 205) being adopted. The apparatus is the most complicated and probably the most efficient for susceptibility measurements at low temperatures. A very ingenious horizontal gas cryostat was constructed, in which the tube containing the sample was cooled by a current of cold gas. Thus the Dewar vessel could be held in a horizontal position and the temperatures obtained were not limited to those of the various possible cooling liquids.

The movable system consisted of a horizontal German silver tube, which carried the thick-walled copper vessel containing the gas at one end. The system was suspended from two pairs of thin wires, each forming a V in the plane perpendicular to the direction of the tube. In this way the motion was confined to one horizontal direction. The system could slide to and fro freely in the cryostat and was placed in the magnetic field in the usual way as required by the "large sample method".

The magnetic force is compensated by a system of solenoids, one of which, in the form of a flat disk, is mounted on the movable system, while two others are fixed in space to right

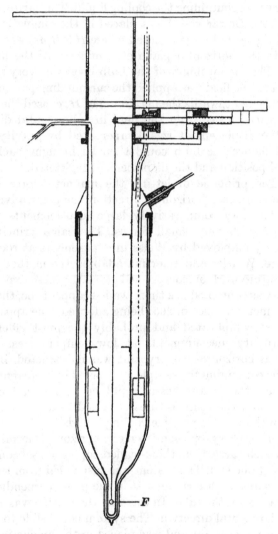

Fig. 2. Apparatus for measuring susceptibility of gases
at low pressures.

and left of the first. A current is transmitted through the three coils in series in such a direction that a slight displacement of the central coil brings about a repulsive force from the coil approached and an attractive force from the other. In this way comparatively large forces may be compensated by very weak currents, the displacement being brought back to zero. To every force corresponds a definite current in the solenoids, which can thus be used to measure the force. For details of this very ingenious apparatus we must refer to the original paper of the authors.

(5) In many laboratories large electromagnets are not available, and numerous clever attempts have been made to replace them. In an apparatus constructed by Simon and Aharoni a small solenoid was used as a magnet and immersed in liquid hydrogen to minimise the resistance. A very strong current was turned on for a period of about 1 sec., which was short enough to prevent too much heat from being developed. The sample was suspended from one arm of a balance *in vacuo* and a counter-weight from the other. The ballistic displacement of the sample could be observed with the help of a small mirror on the balance. By moving an additional solenoid, a position could be found for which no displacement occurred, and from this position the susceptibility of the sample could be determined.

(6) At very low temperatures, when susceptibilities are large, the latter are now frequently measured by an inductance method, which has the great advantage of avoiding movable systems. The essential features of this method are two coils surrounding the specimen in separate circuits. The induction in one coil is measured when a current is turned on or off in the other. The principle is capable of extensive variation depending on the accuracy desired and the conditions of the experiment.

A number of practical difficulties complicate low-temperature magnetic apparatus. In the first place the susceptibility of the medium surrounding the sample must be taken into account. The medium must be carefully freed from paramagnetic substances, especially from oxygen. Thus if liquid nitrogen is used as a cooling agent, great care must be taken

to remove the slightest traces of oxygen and to prevent condensation of oxygen from the atmosphere. Moreover, if the
force is to be determined it is unprofitable to immerse the
specimen in a liquid, as convection currents produce an
irregular motion. In any case boiling must be prevented, as
the bubbles cause the sample to oscillate and thus make
accurate measurements impossible. When a liquid cooling
agent is used, it is customary to enclose the sample in an
additional tube filled with hydrogen or helium or some diamagnetic gas to produce the necessary thermal contact.
A correction must be applied for the susceptibility of the
medium, whether the latter be diamagnetic or paramagnetic.
The maintenance of constant temperature throughout the
length of a large specimen also requires consideration, and
thus it is not surprising that the resulting apparatus is rather
intricate. Apart from all the cryogenic difficulties, the accurate
determination of magnetic susceptibilities as such makes
considerable demands on the skill of the experimenter.

### III. II. 3.  *The Laws of Curie and Weiss*

Long before the concepts of § 1 had been developed, experimental physicists had studied magnetic phenomena and
discovered the general laws governing the behaviour of paramagnetic bodies. In 1895 Pierre Curie stated, as the result of
numerous experiments, that the susceptibility of paramagnetic gases is inversely proportional to the temperature.
A simple thermodynamical proof of Curie's law was given
shortly afterwards by Langevin.

In the formula

$$\chi = \frac{C}{T},\qquad\qquad\dots\dots(1)$$

$C$, which is known as Curie's constant, should have a definite
value for each gas. Thus formula (1), apart from affirming that
the susceptibility varies inversely as the temperature, contains
the statement that $\chi$ is independent of the field strength.
A number of experiments on oxygen by various authors
appeared to show that this is true for field strengths varying
from 100 to 22,000 gauss at room temperature. Curie himself

studied the relation between $\chi$ and $T$ above room temperature and found that $\chi T$ remains constant up to 450° C. The experiments were carried out at 18 atm. pressure. At 100 atm. Kamerlingh Onnes and Oosterhuis found deviations from Curie's law, which could be explained by the fact that at this pressure the deviations from the state of a perfect gas can no longer be neglected.

Few paramagnetic gases exist, and it was therefore natural that Curie should have tested formula (1) for dilute aqueous solutions, whose properties were already known to be similar to those of perfect gases. He showed that the results of Wiedemann and Plessner on solutions of numerous paramagnetic salts were in agreement with the formula. However, the temperature intervals which could be covered were not very great and were bounded by solidification on the one hand and evaporation on the other. It therefore marked an important step forward when Kamerlingh Onnes suggested that Curie's law might be expected to hold for such crystallised substances in which paramagnetic ions were heavily diluted with diamagnetic neighbours. It was, in fact, this proposal that brought magnetism within the reach of low-temperature physics. For at low temperatures a given absolute temperature interval is obviously far more effective for verifying such a law than at high temperatures.

Kamerlingh Onnes and Oosterhuis determined the susceptibility of ferrous ammonium alum, which crystallises with twenty-four molecules of water, between 14·7° and 290° K. The results are shown in Table XX.

Table XX

| $T$ | 290·0 | 169·6 | 77·3 | 64·6 | 20·4 | 17·9 | 14·7 |
|---|---|---|---|---|---|---|---|
| $\chi 10^6$ | 30·4 | 51·8 | 114·7 | 137·0 | 432·0 | 492·0 | 598·0 |
| $\chi T 10^5$ | 882·0 | 879·0 | 887·0 | 885·0 | 881·0 | 881·0 | 879·0 |

In this very considerable temperature interval, in which $\chi$ varies from 30 to 600 . $10^{-6}$, $C = \chi T$ is seen to be constant to $\pm \frac{1}{2}$ per cent. A similar corroboration of Curie's law was found by Jackson for ammonium manganese sulphate in the same interval of temperature and by Kamerlingh Onnes, Perrier and Oosterhuis for gadolinium sulphate. Both these salts crystallise with a large number of water molecules and the paramagnetic manganese and gadolinium ions are separated from one another by numerous diamagnetic ions and atoms.

According to formula (1) $1/\chi$, plotted as a function of absolute temperature, gives a straight line passing through the origin. Now when experimental research was extended to other paramagnetic bodies, which had little or no analogy to perfect gases, it became apparent that in many cases $1/\chi$ was still a linear function of temperature. The only difference in the magnetic behaviour of these substances as against perfect gases and analogous bodies was that the straight line marking the relation between $1/\chi$ and $T$ no longer passed through the origin, but could be produced to intersect the temperature axis at some point above or below absolute zero. Examples for this type of susceptibility-temperature relation are compressed and liquid oxygen, a number of concentrated aqueous solutions and many crystallised salts. The equation describing this relation may be written in the form

$$\chi = \frac{C}{T-\Theta} \quad \text{or} \quad \frac{1}{\chi} = \frac{T}{C} - n. \qquad \ldots\ldots(2)$$

$\Theta$, which marks the point where the straight line $1/\chi$ intersects the temperature axis, is called the Curie point of the substance. In the early stages of research $\Theta$ was found by extrapolating results obtained at temperatures distant from the Curie point, even in cases when the latter was shown to be positive.

Table XXI illustrates the validity of equation (2) for solid anhydrous $CrCl_3$ between $288°$ and $136°K$. Table XXII brings a number of substances that follow this law over a large interval of temperature with their Curie points.

The discrepancies between the results of various authors in Table XXII very probably exceed the limits of error and seem to show that these salts are sensitive to slight impurities, moisture, etc. We see that all the sulphates have negative $\Theta$'s and all the chlorides positive, but apart from this the $\Theta$'s appear to be more or less evenly spread from $-80°$ to $+80°K$.

A plausible interpretation of formula (2) has been suggested

## Table XXI. $CrCl_3$

| $T$ | $10^6\chi$ | $10^5\chi\,(T-32{\cdot}5)$ | |
|---|---|---|---|
| 288·0 | 41·0 | 1048 | |
| 285·9 | 40·6 | 1028 | The values of Table XXI |
| 285·0 | 40·0 | 1010 | are points taken direct |
| 248·3 | 47·8 | 1031 | from Woltjer's Leiden |
| 239·2 | 49·3 | 1018 | measurements. They give |
| 224·3 | 54·0 | 1035 | an idea of the degree of |
| 211·1 | 56·8 | 1015 | accuracy obtained and in- |
| 199·2 | 60·7 | 1013 | cidentally show that at |
| 199·1 | 62·2 | 1036 | 64·2° K. equation (2) evi- |
| 170·7 | 74·0 | 1023 | dently no longer holds for |
| 153·1 | 85·0 | 1025 | $CrCl_3$ |
| 136·1 | 99·6 | 1032 | |
| 64·2 | 294·7 | 934 | |

## Table XXII

| Substance | $\Theta$ | Authors |
|---|---|---|
| $Fe_2(SO_4)_3$ | − 79·5* | Theodorides |
|  | − 73·0 | Ishiwara |
| $CuSO_4$ | − 72·0 | Ishiwara |
| $CoSO_4$ | − 29·9* | Theodorides |
| $MnSO_4$ | − 19·0* | Theodorides |
| $FeSO_4$ | − 16·0 | Ishiwara |
| $MnCl_2$ | + 3·1* | Theodorides |
|  | +21·0 | Ishiwara |
| $COCl_2$ | +47·2* | Theodorides |
|  | +33·0 | Ishiwara |
|  | +20·0 | Woltjer |
| $FeCl_2$ | +20·4 | Woltjer and Wiersma |
| $CrCl_3$ | +44·2 | Ishiwara |
|  | +32·5 | Woltjer |
| $NiCl_2$ | +70·0† | Ishiwara |
|  | +94·0‡ | Ishiwara |
|  | +77·6* | Theodorides |
|  | +67·0 | Woltjer |

\* Determinations above 0° C.     † Below 220° K.     ‡ Above 220° K.

by Weiss. It is clear that in cases when the paramagnetic ions or atoms are not under the conditions of a perfect gas, the influence of neighbouring particles should become noticeable. The exact nature of the forces which thus come into play is still somewhat obscure. Weiss assumed that the *effect* of these forces could be identified with that of a magnetic field proportional to the magnetisation and acting in the same direction though not necessarily in the same sense. This *molecular field* can thus be expressed by the formula $H_m = n\sigma$, where $n$ is a characteristic constant of the substance. Curie's law, as defined by equation (1), thus assumes the form

$$\sigma = C \frac{H + H_m}{T} = C \frac{H + n\sigma}{T},$$

which makes the susceptibility

$$\chi = \frac{C}{T - nC}.$$

This equation becomes identical with (2) if $\Theta = nC$. Since $\Theta$ may be positive or negative the same must be true of $n$, so that the molecular field may act in the same sense as $H$ or in the opposite sense, thus in one case increasing, in the other decreasing, the effect of the external field.

Equation (2) is frequently described as Weiss' law. Heisenberg has given a quantum-mechanical interpretation of Weiss' picture with the help of the exchange effect between neighbouring ions.

The ferromagnetic metals Fe, Co and Ni and the Heussler alloys have susceptibilities in excellent agreement with equation (2) at temperatures above their Curie points, behaving like genuine paramagnetic substances. It therefore appeared probable that paramagnetic salts with positive values of $\Theta$ might become ferromagnetic below their Curie points. The principal criteria for ferromagnetism were considered to be saturation, spontaneous magnetism and hysteresis; but a susceptibility depending in any way on field strength was considered a sign of approaching ferromagnetism.

However, there was some reason to doubt the physical significance of these positive Curie points. The susceptibilities

of numerous substances, that followed equation (2) very
accurately down to the temperatures of liquid air, began to
show deviations below this region, even in some cases where

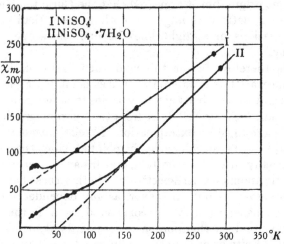

Fig. 3. Cryomagnetic anomaly in nickel sulphate.

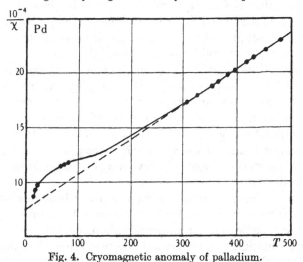

Fig. 4. Cryomagnetic anomaly of palladium.

the Curie point was still far away. The susceptibility temperature curves were no longer straight lines but began to
form hooks, as shown in figs. 3 and 4. This effect was termed

the *cryomagnetic anomaly*. It was not known whether all bodies exhibited this anomaly at sufficiently low temperatures or even whether the effects observed in various bodies were of the same origin, but it appeared that the Curie temperatures of paramagnetic salts might be nothing more than pointless extrapolations of straight lines beyond the scope of their validity.

It is therefore better to avoid such extrapolations and introduce a "ferromagnetical" Curie point $\Theta_F$ as the temperature at which a body actually begins to show ferromagnetic properties. $\Theta_F$ may well differ from the "paramagnetical" $\Theta_P$ which is obtained by extrapolation of Weiss' formula. It may for simplicity be considered as the point at which the curve for $1/\chi$ actually intersects the temperature axis after any cryomagnetic anomalies, irrespective of whether these have as yet been observed or not. However, as any dependence of $\chi$ on the field strength may be considered as a ferromagnetic property, and as this does not necessarily demand that $\chi$ become infinite or even exceedingly great, $1/\chi$ may well have a finite value at $\Theta_F$. $\Theta_F$ may in principle be greater or less than $\Theta_P$; it may even be negative when $\Theta_P$ is positive. It is natural to suppose that a cryomagnetic anomaly that takes $1/\chi$ below the straight line of Weiss' formula will lead to $\Theta_F > \Theta_P$ and vice versa, but even this is not self-evident. The introduction of $\Theta_F$ can be justified only by experiment. It must be determined by experiment whether any ferromagnetic phenomena occur at all and especially whether at sufficiently low temperatures $\chi$ depends on $H$; whether this begins sharply at a definite temperature, and whether the position of $\Theta_F$, if it exists, is connected with the cryomagnetic anomalies in the way indicated.

Some of these experiments have already been performed. Kamerlingh Onnes, Woltjer and Wiersma have investigated the susceptibilities of a number of anhydrous chlorides below $\Theta_P$ as a function of temperature and field strength, using the large sample method described in § 2 as (2). Some of their results are shown in fig. 5 *a, b, c, d*.

Down to at least 137·6° K. $NiCl_2$ follows Weiss' law with $\Theta_P = 67°$. At lower temperatures a cryomagnetic anomaly

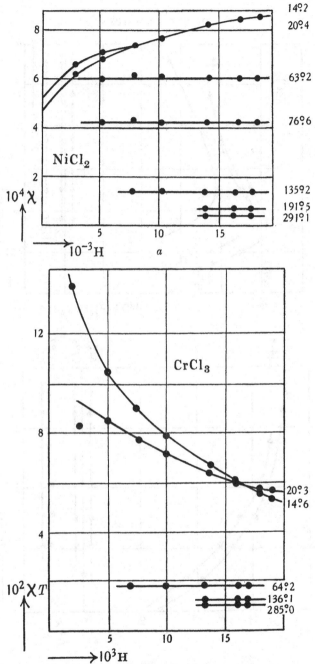

Fig. 5. Magnetic susceptibilities of anhydrous chlorides.
(a) NiCl₂, (b) CrCl₃.

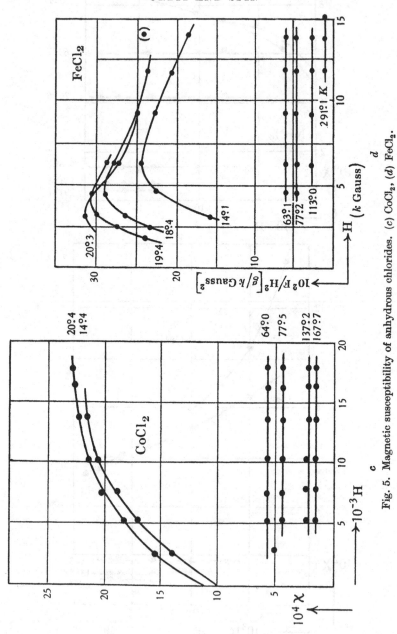

Fig. 5. Magnetic susceptibility of anhydrous chlorides. (c) CoCl$_2$, (d) FeCl$_2$.

occurs, as a result of which $1/\chi$ comes to lie *above* Weiss'
straight line. For at 77·6° $\chi T$ is only half as great as at higher
temperatures, i.e. $1/\chi$ is twice as great as might be inferred
from extrapolation. Moreover, at 63·9°, i.e. below $\Theta_P$, $\chi$ does
not yet depend on the field strength. Yet at hydrogen temperatures $\chi$ does depend on the field strength, and therefore
this effect must begin somewhere below $\Theta_P$. However, as no

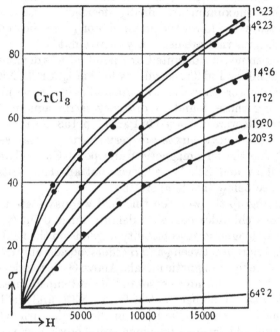

Fig. 6. Magnetisation of $CrCl_3$.

determinations were made at intermediate temperatures, we
cannot conclude from these experiments that there is a well-
defined limiting temperature $\Theta_F < \Theta_P$. This conclusion can,
however, be drawn from other experiments which we shall
treat shortly. Let us first see to what degree these chlorides
actually show ferromagnetic properties at hydrogen temperatures. In $NiCl_2$ and $CoCl_2$ $\chi$ increases with the field strength,
but in $CrCl_3$ it decreases, giving a relation between $\sigma$ and $H$
as shown in fig. 6.

These curves point to approaching saturation, and in this respect $CrCl_3$ bears some resemblance to a ferromagnetic substance. However, only the incipient stages appear, and at values of $H$ that are very much higher than those that give almost full saturation in iron and nickel. No spontaneous magnetisation or hysteresis could be detected. $FeCl_2$ shows a behaviour intermediate between $CrCl_3$ and $NiCl_2$. At hydrogen temperatures $\chi$ first increases with rising $H$, passes through a maximum and finally decreases. The lower the temperature, the lower the maximum of $\chi$ and the higher the field strengths that are necessary to attain it.

The existence of a definite Curie point $\Theta_F$ was demonstrated by Schubnikow and Trapeznikova for $FeCl_2$, $CrCl_3$, $NiCl_2$ and $CoCl_2$ which were shown to have anomalies in the specific heats resembling those found in ordinary ferromagnetic bodies. These results were discussed in Part II, p. 158. As the maxima in the specific-heat curves are very sharp, we may say that they define the ferromagnetic Curie point $\Theta_F$, and recently Schubnikow and Schalyt showed that all these salts show remanence below this temperature.

Summing up the work on this field we may state that the anhydrous chlorides possess a definite Curie point $\Theta_F$ below which the magnetic susceptibility depends on the field. However, the relation between $\chi$ and $H$ differs considerably from that found in the ferromagnetic metals. Above $\Theta_F$, $\chi$ is independent of $H$, but Weiss' formula ceases to hold at temperatures which are considerably higher. $FeCl_2$, $CoCl_2$, $NiCl_2$ and $CrCl_3$ show anomalies in their specific heats in the neighbourhood of $\Theta_F$ of the type observed in ferromagnetic bodies near the Curie point. $\Theta_F$ does not coincide with the $\Theta_P$ obtained by extrapolating Weiss' formula.

According to Weiss ferromagnetic phenomena may be understood by supposing the existence of a term in the molecular field which is independent of the field strength. This leads to spontaneous magnetisation in "elementary groups" of atoms. In the absence of a field this magnetisation is not as a rule noticeable, as the groups are orientated arbitrarily. However, a very small field strength suffices to orient these groups, so that the typical magnetisation curves of ferromagnetic bodies may be deduced. This theory accounts satisfactorily for the relation between magnetisation and temperature in the neighbourhood of the Curie point and for the form of the specific-heat curve. With the help of Heisenberg's quantum mechanical exchange effect the same formulae are obtained as with Weiss' phenomenological concepts. To explain the differences in behaviour observed between the paramagnetic salts and the ferromagnetic metals, Landau suggested that within the layers occupied by the paramagnetic ions in these crystals the interaction forces are such as to orient the magnetic moments of these ions in one direction, whereas the weaker forces acting from layer to layer serve to orient neighbouring layers in opposite directions. Thus no spontaneous magnetisation occurs, as large fields are needed to counteract this orientation. The formulae deduced from this concept agree fairly well with the experimental results and show further that the temperature $\Theta_F$ at which $\chi$ may be expected to begin to depend on $H$ need not coincide with the $\Theta_P$ obtained by extrapolating Weiss' equation.

### III. II. 4.  *Oxygen and Nitric Oxide*

The two paramagnetic gases oxygen and nitric oxide are particularly suited to exemplify the methods and calculations discussed in the preceding paragraphs. Practically all the types of apparatus described above have been employed in determining the susceptibility of oxygen in its various forms. The properties of the two gases are in many respects diametrically opposed to one another. The ground level of NO is a $^2\Pi$-term, i.e. a doublet. The two levels of the doublet $^2\Pi_{\frac{3}{2}}$ and $^2\Pi_{\frac{1}{2}}$ are regular, i.e. $^2\Pi_{\frac{3}{2}}$ is higher than $^2\Pi_{\frac{1}{2}}$, and the energy difference $\Delta E$ between these two terms is of the order of magnitude of $kT$ at room temperature. Thus none of the approximations valid for $\Delta E \gg kT$ or $\Delta E \ll kT$ can here be used except to determine the limiting values of $\chi$ for high and low temperatures. For these cases formulae (9) and (10), respectively, on p. 204, should hold. Inserting $S = \frac{1}{2}$ and $\lambda = 1$ in (9) and $\sigma = -\frac{1}{2}$ and $\lambda = 1$ in (10) we obtain $\chi = (2\mu_0)^2/3k$ for very high temperatures and $\chi = 0$ for $T \to 0$. The relation between

$\chi$ and $T$ has been calculated accurately by van Vleck. The result can be expressed in the form

$$\chi = \frac{N\Theta^2}{3kT} \text{ with } \Theta^2 = 4\mu_0^2 \frac{1 - e^{-x} + xe^{-x}}{x + xe^{-x}} \text{ and } x = \frac{h\Delta V}{kT}.$$

Here $\Delta V$ is the difference between the two $^2\Pi$ levels and $x = 173\cdot2/T$.

Van Vleck's calculations were tested by Wiersma, de Haas and Capel in 1930, the method denoted as (3) and described on p. 208 being employed.

The results are shown in Table XXIII, in which the experimental values of $\Theta/\mu_0$ are compared with those computed by van Vleck. The agreement is seen to be fairly good, very slight deviations occurring at low temperatures, which are probably due to secondary causes.

Table XXIII.  *Susceptibility of* NO

| $T^\circ$ K. | $\chi.10^6$ | $C = \chi T . 10^3$ | $C$ mol. | $\Theta/\mu_0$ | $\Theta/\mu_0$ according to v. Vleck | Differences per 1000 |
|---|---|---|---|---|---|---|
| 292·10 | 49·07 | 14·336 | 0·4302 | 1·852 | 1·838 | +7 |
| 238·40 | 56·42 | 13·450 | 0·4036 | 1·794 | 1·794 | . |
| 226·27 | 58·87 | 13·320 | 0·3997 | 1·785 | 1·783 | +1 |
| 214·45 | 61·21 | 13·126 | 0·3939 | 1·772 | 1·770 | +1 |
| 203·80 | 63·16 | 12·873 | 0·3863 | 1·755 | 1·756 | −1 |
| 170·84 | 71·45 | 12·207 | 0·3663 | 1·709 | 1·708 | +1 |
| 165·41 | 72·26 | 11·953 | 0·3587 | 1·691 | 1·694 | −2 |
| 158·75 | 73·59 | 11·683 | 0·3506 | 1·672 | 1·682 | −6 |
| 146·90 | 77·45 | 11·377 | 0·3414 | 1·650 | 1·655 | −3 |
| 138·29 | 79·11 | 10·940 | 0·3283 | 1·618 | 1·630 | −8 |
| 132·51 | 81·45 | 10·794 | 0·3239 | 1·607 | 1·616 | −6 |
| 112·77 | 87·32 | 9·847 | 0·2955 | 1·535 | 1·547 | −8 |

Whereas nitric oxide is a classical example of a paramagnetic gas, the susceptibility of which does not obey Curie's law, oxygen was long considered to be a veritable "Curie gas", as indeed theoretical considerations would lead us to suppose.

The ground level of oxygen is a $^3\Sigma$ term of the type denoted by (b) on p. 204. It is a very narrow triplet, the distances between the three components being negligible above $1^\circ$ K.

We should thus expect equation (9), p. 204, to hold with $S = 1$. This would give

$$\chi = \frac{8}{3} \cdot \frac{\mu_0^2}{kT},$$

where $\mu_0$ is Bohr's magneton.

In chap. II, § 3, we remarked that early experiments at room temperature and above were in good agreement with Curie's law. At low temperature ($-130°$) and pressures of 100 atm. slight deviations were observed that could be attributed to experimental errors or to the fact that under these conditions oxygen may no longer be considered as a perfect gas. The supposed relation between the susceptibility and density of oxygen was studied in a series of very beautiful experiments by Kamerlingh Onnes and Perrier on liquid mixtures of oxygen and nitrogen, an apparatus of type (2) being employed. The fact that the susceptibility of pure liquid oxygen did not follow Curie's law could be plausibly explained by assuming that the susceptibility depends on the density, which varies considerably with the temperature. To obtain oxygen of the same density at various temperatures Onnes and Perrier added definite quantities of liquid nitrogen as a diluting agent. In this way they obtained a series of parallel straight lines giving the relation between $1/\chi$ and $T$ for various densities (see fig. 7). The fact that these lines are parallel shows that the Curie constant $C$ is independent of the density, whereas the fact that they are straight intimates that Weiss' law is valid:

$$\chi = \frac{C}{T - \Theta}.$$

$\Theta$ could be expressed as $-a\rho$, where $\rho$ is the density and $a$ a characteristic constant. Extrapolation to $\rho = 0$ gives Curie's law in the original form. It thus appeared that in liquid oxygen-nitrogen mixtures we have to do with a "molecular field" proportional to the density of oxygen in the mixture.

The next step was to test this law for pure gaseous oxygen at high pressures and at densities similar to those of the oxygen in the liquid mixtures. This was done with the help of the horizontal apparatus described as type (4) (see p. 209), after type (2) had been discarded as insufficiently sensitive. As little importance was attached to the absolute susceptibilities the results were expressed in the relative form $\chi_T/\chi_{291°}$. In order to demonstrate the deviations from Curie's law $\chi_{291°}/\chi_T - T/291°$ was plotted as a function of $T$ (fig. 8). If the original form of Curie's law were valid, this expression should be zero for all temperatures and densities. If Weiss' law with $\Theta = -a\rho$ holds, the experimental points obtained at different densities should

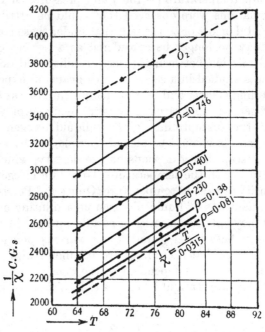

Fig. 7. Susceptibility of liquid oxygen-nitrogen mixtures.

be grouped along a series of straight lines each valid for a definite density. From the figure it is clear that neither of these assumptions is true. The points are grouped about a curve that is quite definitely not straight. Moreover, the density appears to have very little influence at all on $\Theta$; evidently the squares, circles and triangles are scattered at random. The figure shows, however, that at temperatures above about $-175°$ C. the deviations from Curie's law are slight and unsystematic. But we find that in the formula

$\chi T = C$, $C$ depends on the density. The following values of $C$ were found for different densities:

| $\rho = 0.443$ | 0·32 | 0·152 | 0·00 |
|---|---|---|---|
| $10^2 C = 2.86$ | 2·93 | 3·05 | 3·15 |

This would mean that the susceptibility of the oxygen molecule decreases when the density is increased.

These results were very surprising and hard to reconcile with those obtained with the liquid mixtures, especially the

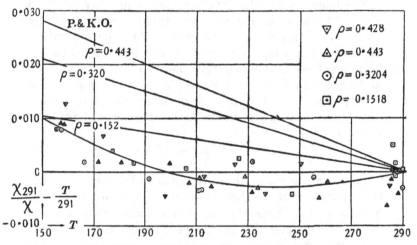

Fig. 8. Susceptibility of oxygen at high pressure.

fact that the deviations from Curie's law appeared to be independent of density. It was thus of interest to ascertain whether the same deviations occur at normal pressure. For these experiments the same apparatus of type (3) was used that had been employed in the work on NO. The results, which up till now are the latest to hand, are as follows. Between 78° and 249° K., the temperature interval covered, $\chi T$ is definitely not a constant. Tolerable agreement is obtained with a Weiss equation $\chi = C/(T + 1·7)$, but the agreement is probably better if Curie's form is discarded and we write $\chi = (C/T) + \alpha$ ($\alpha = 1·8 . 10^{-6}$). This would lead us to assume the existence of a small paramagnetic term independent of the temperature, a possibility which had already been predicated by van Vleck as due to the high frequency elements of the magnetic moments.

We may therefore take it that the susceptibility of oxygen at low pressures is in agreement with existing theory. However, at high pressures and in the liquid state some additional assumption must be made to explain the experimental facts here mentioned as well as a number of other more or less anomalous phenomena connected with oxygen in its various aggregate states.

The dependence of the Curie point $\Theta$ of liquid oxygen on the density, as shown by the work on oxygen-nitrogen mixtures, leads to the conception of a molecular field which depends on the density of the liquid. Lewis suggested that this field might be a sign of association of oxygen molecules. From the results of Kamerlingh Onnes and Perrier he derived values for the equilibrium constant $\rho_{O_2}^2/\rho_{O_4} = K_\rho$ of the reaction $2O_2 \rightleftharpoons O_4$.

By extrapolation to the gaseous state he obtained the formula

$$- R \ln K_\rho = - \frac{12}{T} - 2 \ln T + 8 \cdot 67.$$

This idea was taken up by Wiersma and Gorter to explain the dependence of the Curie constant $C$ on density in gaseous oxygen at high pressure. As $\chi$ is approximately a linear function of $\rho$, association to $O_4$ seemed probable. Assuming $O_4$ to be diamagnetic Wiersma and Gorter determined $K_\rho$. The following values were obtained:

| $T$ | $K_\rho$ gm./cm.$^3$ | Authors |
|------|------|------|
| 155·0 | 3·3 | Wiersma and Gorter |
| 155·0 | 2·2 | Lewis' formula |
| 290·0 | 4·3 | Wiersma and Gorter |
| 290·0 | 4·0 | Lewis' formula |
| 64·2 | 0·92 | Lewis (liquid) |
| 77·4 | 1·10 | Lewis (liquid) |

The order of magnitude is the same according to Lewis and also Wiersma and Gorter. Not much more can be expected. Nevertheless, these results are in themselves insufficient to justify the hypothesis of association, more especially as the dissociation energy of $O_4$ is so small that the attractive forces must be of the order of magnitude of the van der Waals forces. However, several other phenomena seem to point in the same direction. Liquid and compressed gaseous oxygen possess an absorption spectrum in the visible and near ultra-violet region that is not observed in the atmosphere. Several intense bands in the red, yellow and blue were already observed by Olszewsky in the liquid. Steiner and Salow observed these bands in gaseous oxygen at room temperatures at pressures up to 600 atm. and found the intensity to increase quadratically with the pressure, whereas the well-known infra-red atmospheric absorption bands are connected with pressure in a linear relationship. This could be explained very plausibly by assuming the spectrum to be due to $O_4$. Shortly afterwards Guillen studied the absorption spectrum of liquid oxygen as a function of density. Just as in the experiment of Kamerlingh Onnes and Perrier the liquid oxygen was diluted with nitrogen to vary the density. The result showed the intensity of absorption to be proportional to the square of the density, as in the case of compressed gaseous oxygen. Moreover, assuming the spectrum to be due to $O_4$, the reaction constant $K_\rho$ could be determined from the relative intensities of the spectrum. A figure was obtained, which was in good agreement with those determined from the magnetic measurements. Finally Ellis and Kneser succeeded in describing all the visible and near ultra-violet bands of oxygen with the help of the three known $O_2$ levels $^3\Sigma$, $^1\Delta$ and $^1\Sigma$, by supposing double transitions to occur from the $^3\Sigma$ ground level to the other two, in which the frequencies of the two transitions are merely added together. Thus if $(^1\Delta)$ and $(^1\Sigma)$ signify the frequencies of the transitions $^3\Sigma - {}^1\Delta$ and $^3\Sigma - {}^1\Sigma$ respectively, the entire spectrum can be covered by a formula $\nu = m\ (^1\Delta) + n\ (^1\Sigma) + \sigma\nu_v$, where $\nu_v$ is a vibrational frequency corresponding to $O_2$ and $m$, $n$ can be 0, 1 or 2. Although this explanation has several theoretical difficulties, it covers the case too well to be simply discarded. According to experiments by Prikhotjko, Ruhemann and Fedoritenko, these so-called $O_4$ bands become very strong indeed in the lower temperature $\alpha$- and $\beta$-modifications of solid oxygen (see Part II, p. 117), whereas in the upper $\gamma$-form they are of the same intensity as in the liquid. This corresponds with old experiments of Kamerlingh Onnes on the magnetic susceptibility of solid oxygen. On crystallisation to the $\gamma$-modification only a slight decrease of susceptibility occurs, but at lower temperatures at which $\beta$- and $\alpha$-oxygen are stable $\chi$ becomes very small indeed and decreases with falling temperature. If we suppose $O_4$ to be diamagnetic, the increasing intensity of the $O_4$-spectrum and the simultaneous decrease in the susceptibility may be taken as a measure for the rising concentration of $O_4$.

### III. II. 5. *Saturation, Langevin's Formula and the Faraday Effect*

The old interpretation of Curie's law considered the atoms or molecules of a paramagnetic gas as magnetic dipoles, each possessing identical magnetic moments $\mu$. These dipoles would tend to be orientated in a homogeneous magnetic field, thermal agitation impeding this orientation. On the basis of this model it is natural that at a given field strength the degree of orientation, and thus the mean magnetic moment of a certain quantity of gas, should increase with falling temperature, whereas at a given temperature it should increase when the field strength is raised. However, in a sufficiently strong field, when the dipoles are almost completely orientated already, an increase in the field strength can have little influence on the mean magnetic moment. Under these circumstances the magnetisation of the gas should approach saturation.

The formula deduced on the basis of this classical model by Langevin, connecting the magnetic moment $\sigma$ of a gramme-molecule of gas with field strength and temperature, is generally written

$$\sigma = \sigma_0 \left( \coth \xi - 1/\xi \right), \qquad \ldots\ldots(1)$$

where $\sigma_0$ is the product of the atomic magnetic moment $\mu$ and Avogadro's number and $\xi = \mu H/kT$. Expressed as a series this formula may be written

$$\frac{\sigma}{\sigma_0} = \frac{\xi}{3} - \tfrac{1}{45}\xi^3 + \tfrac{2}{945}\xi^5 + \ldots \qquad \ldots\ldots(2)$$

For small values of $\xi$ we may break off the series after the first term and obtain

$$\sigma = \frac{\mu H \sigma_0}{3kT} \quad \text{or} \quad \sigma = \frac{\sigma_0^2 H}{3RT},$$

which gives

$$\chi = \frac{\sigma_0^2}{3RT}.$$

Comparing this with equation (1) of § 3, we obtain for small values of $\xi$ Curie's law with the constant

$$C = \frac{\sigma_0^2}{3R},$$

$\chi$ being now defined as the magnetic moment of one gramme-molecule divided by the field strength. We saw in equation (8), p. 203, that in cases when only the ground level need be taken into account quantum theory yields Curie's law with

$$C = \frac{J(J+1)g^2\mu_0^2}{3k}.$$

Expressed in units of one gramme-molecule this may be written

$$C = \frac{J(J+1)g^2(\mu_0 N)^2}{3R},$$

so that the two formulae become identical if

$$\sigma_0 = \sqrt{J(J+1)}\, g\mu_0 N.$$

At higher values of $\xi$ Langevin's formula leads to deviations from Curie's law and finally to saturation. However, the magnetic moments of all paramagnetic substances are so small that the law should hold under almost all conditions obtainable in the laboratory. To obtain large values of $\xi$ we may either raise the field strength or lower the temperature. Naturally low-temperature physics is interested primarily in the latter possibility, but it will be as well to ascertain first what can be done by employing large fields. Take, for example, a dilute solution of some manganese salt, the ions of which have a particularly large magnetic moment: $\sigma_0 = 32,800$ per gramme-atom. Then at room temperature

$$\xi = \frac{\mu_0 H}{kT} = \frac{\sigma_0 H}{RT} = \frac{32,800}{8\cdot315.10^7} \cdot \frac{H}{288}.$$

Taking $H = 30{,}000$ gauss, which is about the maximum field strength of a large laboratory magnet, we obtain

$$\xi = 0 \cdot 0411.$$

From Langevin's formula we may conclude that for $\xi = 0 \cdot 12$ the deviation from Curie's law is still less than $0 \cdot 1$ per cent. We thus see that if Langevin's formula comes anywhere near the facts, saturation phenomena can hardly be observed at room temperature. It is evident that very high fields combined with extremely low temperatures are needed to follow up Langevin's curve to values approaching saturation.

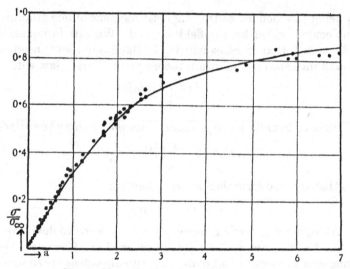

Fig. 9. Magnetisation of gadolinium sulphate.

The crucial test of Langevin's formula was carried out by Kamerlingh Onnes and Woltjer in 1923 on gadolinium sulphate $(Gd_2(SO_4)_3 . 8H_2O)$, employing field strengths up to 22,000 gauss and the lowest temperatures then obtainable with the help of liquid helium. The results, which are given in fig. 9, show that not only could considerable deviations from Curie's law be determined, but that Langevin's curve could be covered to about 85 per cent. of saturation.

At a first glance the agreement between theory and experiment appears surprisingly good. The points lie very near the theoretical curve and the deviations were at the time considered to be due to experimental error or to the fact that the conditions of a perfect gas are not completely realised in this dilute solid solution of gadolinium ions. A close inspection discloses the fact that these deviations are indeed systematic. The experimental points lie slightly above the curve at lower values of $\xi$ and slightly below when $\xi$ is greater. In actual fact the approach to saturation is rather steeper than would follow from Langevin's formula. Considering that this formula is based on purely classical principles this is not very surprising. In 1926 Debye pointed out that a literal application of Langevin's equation would lead to contradictions with Nernst's theorem, which demanded a steeper approach to saturation. We shall refer to this point again in chap. III. In numerous papers, and especially in the work of van Leeuwen, it was finally shown that Langevin's model of magnetic dipoles was not permissible in the quantum theory. The quantum mechanical calculations were ultimately carried out by van Vleck and Niessen. No general formula can be given to meet all cases, as the exact form of the saturation curve depends on the nature of all the lower levels, but it can be shown that the approach to saturation should indeed be steeper than according to Langevin's formula.

Gadolinium sulphate is the only substance the magnetisation of which has been measured up to sufficiently high values of $H/T$ so that saturation can be determined. However, research on a quite different field has yielded fresh results on paramagnetic saturation. It is a well-known fact that when a beam of plane-polarised light is passed through a liquid in a magnetic field, the plane of polarisation is turned through an angle which, for a given liquid, is proportional to the field strength. This is known as the Faraday effect. The effect is not confined to liquids but is observed in a number of transparent solids. Henri Becquerel showed as early as 1877 that the magnetic rotation is connected with the magnetic properties of the particles in the liquid or solid. In 1884 Kundt discovered that thin transparent films of ferromagnetic metals possess a very strong rotational power, which is *not* proportional to the field strength but shows saturation. We may therefore say that the Faraday effect is very closely connected with the *magnetisation* of the medium. As a practical working hypothesis it was natural to assume that the rotational power $\rho$ is proportional to the magnetisation $\sigma$ and is thus connected with $H/T$ by a formula of the Langevin type.

Jean Becquerel found that a number of crystallised compounds of the rare earths possess a considerable rotational power. He pointed out that this is connected with the absorption bands of these crystals, the absorption differing for two beams of light circularly polarised in different directions. Jean Becquerel conducted a number of experiments at Leiden on the magnetic rotation of crystals of the rare earths at low temperatures as a function of frequency, temperature and field strength. As was to be expected, $\rho$ increased very rapidly as the temperature was reduced.

Becquerel worked at the temperatures of liquid nitrogen, liquid hydrogen and liquid helium, in some cases as low as $1.3°$ K. At the lowest temperatures he used a glass cryostat consisting of three concentric Dewar vessels, which is one of the masterpieces of the Leiden workshop. The inner tube of the innermost Dewar, which contained liquid helium, was 5 mm. in diameter, and the outer tube of the outside vessel was only 15 mm. wide. The intermediate vessel contained liquid hydrogen. Thus the light had to pass through twelve glass walls apart from the crystal; yet it is claimed that the optical conditions were good.

Several forms of optical apparatus were employed, one of which may be briefly described. It was designed to enable large and rapidly varying angles of rotation to be determined quickly and fairly accurately. The crystal was mounted in the innermost Dewar vessel with its optical axis parallel to the horizontal magnetic field. The poles of the magnet are pierced with narrow holes and a beam from an arc lamp is plane polarised and directed through the poles, Dewar vessels and crystal and made to fall on the slit of a spectrograph through an analyser, consisting of a rhombohedron of Iceland spar. This analyser evolves two spectra side by side of perpendicularly plane-polarised light. In the absence of a magnetic field the analyser is so adjusted that only one spectrum appears.

On closing the current of the magnet the Faraday effect, which depends on the wave-length, causes the spectrum to break up into light and dark narrow bands. A dark band on one spectrum corresponds to a light band on the other. When the field strength is gradually raised or lowered, the dark bands move across the field of vision. A fine absorption line of the crystal or an emission line of an iron arc, projected on the slit, is used as a mark and the field strength registered every time a black band passes through this line.

Most of Becquerel's experiments were carried out on tysonite, a crystal consisting of (La, Ce) $F_3$ with traces of Nd and Pr. It could be shown that only the Ce ions are effective. Tysonite shows the strongest rotational power. The large Leiden electromagnet was used, giving field strengths up to 27,000 gauss. The thickness of the crystals was between 1 and 2 mm.

In order to convey an impression of the enormous rotational power of tysonite at low temperatures, some of Becquerel's results are shown in Table XXIV for a temperature of $1 \cdot 32°$ K. $\rho$ is reduced to a thickness of 1 mm. and given in degrees. The values of $H$ are corrected; in actual fact they cannot be determined so accurately. These results correspond to the green mercury line $\lambda = 5460 \cdot 7$ Å.

Table XXIV.  *Rotational power of tysonite at* $1 \cdot 32°$ K.

| $H$ (gauss) | $\rho$ (degrees) | $H$ (gauss) | $\rho$ (degrees) |
|---|---|---|---|
| 5303 | 459·17 | 21717 | 1397·55 |
| 9065 | 755·16 | 24328 | 1467·85 |
| 11961 | 955·42 | 25784 | 1497·84 |
| 17622 | 1248·19 | 26997 | 1520·06 |

At helium temperatures the linear relation between $\rho$ and $H$ ceases to hold above 5000 gauss. Thereafter the curve approaches saturation and at $1 \cdot 32°$ K. and 27,000 gauss 85 per cent. of saturation is attained. The curve is very similar to that of the magnetisation of gadolinium sulphate shown in fig. 9, p. 232. The values of $\rho$ for full saturation depend on frequency and temperature. After his first experiments Becquerel reduced all his results to one curve, in which the percentage saturation was plotted as a function of $H/T$. It was clear that the approach to saturation was steeper than according to Langevin's formula. A better agreement was obtained with a formula developed by Lenz and Ehrenfest for the magnetisation in one of the principal directions of a crystal. This formula could be written in the form

$$\rho = \rho_\infty \tanh \frac{\mu H}{kT}, \qquad \qquad \ldots\ldots(3)$$

where $\rho_\infty$ depends on $\lambda$ and $T$. Moreover, it could be shown that $\mu$ was almost exactly equal to Bohr's magneton.

However, this was in contradiction to theoretical and experimental results concerning the magnetic moment of the Ce$\cdots$ ion, which had been shown to be responsible for the magnetic rotation. From Gorter's determination of the susceptibility of CeF$_3$ down to hydrogen temperatures, Kramers deduced $\mu = \frac{9}{7}\mu_0$, where $\mu_0$ is Bohr's magneton.

In view of these difficulties Becquerel repeated his measurements on tysonite with greater accuracy, especially as concerned the determination of temperature. It was found that equation (3) did not give an accurate picture of the results and a more complicated formula was developed, into which a "molecular field" of the Weiss type entered.

According to van Vleck and Hebb the magnetic rotation should, in the case of the rare earths, be proportional to the magnetisation. Moreover, Kramers showed that when only the ground level of the rare earth ion is excited a formula of the type

$$\rho = A \tanh \frac{\mu H}{kT} + BH \qquad \qquad \ldots\ldots(4)$$

should hold, where, at sufficiently low temperatures, $A$ and $B$ are functions of the wave-length only. Becquerel was able to confirm this formula in recent experiments on ethylsulphates of rare earths. The magnetic rotation is much smaller in these salts than in minerals like tysonite, but we have the advantage of working with pure substances. In dysprosium- and erbium-ethylsulphate Dy (or Er) $(C_2H_5SO_4)_3 . 9H_2O$ complete magnetic saturation occurs at field strengths between 10 and 20 kilogauss below 4° K. In the case of erbiumethylsulphate we have the particularly simple case of $B = 0$.

# CHAPTER III

## MAGNETIC COOLING

### III. iii. 1.  *The Magneto-Caloric Effect*

We have seen that paramagnetic bodies contain atoms, ions or molecules in degenerate states, and we know that the terms corresponding to these states have multiple weights which lead to additional entropy. From the formulae on p. 195 we may conclude that a statistical weight $P$ in the ground level will produce at low temperatures an additional term $R \log P$ in the entropy. Now under the influence of a magnetic field the degenerate terms are split up into a number of Zeeman components with smaller statistical weights. Thus a magnetic field causes the additional term in the entropy to be diminished.

Now consider a paramagnetic body surrounded by an adiabatic jacket and placed between the poles of an electro-magnet. When the current is turned on the degenerate terms are split up and the magnetic entropy lowered. However, owing to the adiabatic jacket, the total entropy of the body must remain unchanged in this process. Thus the loss of magnetic entropy must be compensated by an increase of entropy from another source. This takes the form of a temperature increase from $T_1$ to $T_2$, the entropy thus gained, $\int_{T_1}^{T_2} \frac{C}{T} dT$, being exactly equal to the loss of magnetic entropy incurred. When the magnetic field is removed, and thus the magnetic entropy increased, a corresponding fall in the temperature is observed. This change of temperature, positive when the field strength is raised and negative when it is lowered, is known as the Magneto-Caloric Effect.

The differential magneto-caloric effect, i.e. the temperature change $dT$ caused by a change of field strength $dH$, can be computed thermodynamically as follows. To produce a change of magnetisation $d\sigma$, a field must perform an amount of work $H d\sigma$. Thus in an adiabatic process the First Law requires that

$$dU - H d\sigma = 0. \qquad \ldots\ldots(1)$$

Taking temperature $T$ and field strength $H$ as independent variables and neglecting possible slight variations in volume and pressure, we may write (1) in the form

$$\frac{\partial U}{\partial H} dH + \frac{\partial U}{\partial T} dT - H \frac{\partial \sigma}{\partial H} dH - H \frac{\partial \sigma}{\partial T} dT = 0$$

or

$$\left( \frac{\partial U}{\partial H} - H \frac{\partial \sigma}{\partial H} \right) dH + \left( \frac{\partial U}{\partial T} - H \frac{\partial \sigma}{\partial T} \right) dT = 0.$$

Now from a well-known thermodynamical relation we may deduce

$$\frac{\partial U}{\partial H} - H \frac{\partial \sigma}{\partial H} = T \frac{\partial \sigma}{\partial T},$$

whereas

$$\frac{\partial U}{\partial T} - H \frac{\partial \sigma}{\partial T} = C_H,$$

the specific heat in a constant field. Thus

$$T \frac{\partial \sigma}{\partial T} dH = - C_H dT$$

and

$$\left( \frac{\partial T}{\partial H} \right)_{s=\text{const.}} = - \frac{T \left( \frac{\partial \sigma}{\partial T} \right)_H}{C_H}. \qquad \ldots\ldots(2)$$

In order to form an idea of the order of magnitude of the magneto-caloric effect, we may apply equation (2) to a paramagnetic gas, say oxygen at room temperature. Here, according to Curie's law,

$$\sigma = \chi H = \frac{CH}{T},$$

if we neglect the slight deviations mentioned in the last chapter, and we may integrate equation (2) from 0 to $H$, obtaining

$$\int_{T_1}^{T_2} T dT = \int_0^H \frac{CH}{C_H} dH$$

or

$$\frac{T_2^2 - T_1^2}{2} = CH^2/2C_H. \qquad \ldots\ldots(3)$$

We shall presently see that $T_2 - T_1$ is very small, so that we may define an average $T$ as $T = (T_2 + T_1)/2$. Equation (3) then becomes

$$T_2 - T_1 = CH^2/2C_H T. \qquad \ldots\ldots(4)$$

Now assuming that $C_H$ is not very different from the specific heat $C_p$ at constant pressure, we obtain for oxygen

$C_H = 0.156$ cal. per degree and gramme
$= 0.156 \times 4.18 \times 10^7$ ergs per degree and gramme.

The Curie constant $C = 0.0316$, $T = 293°$, which makes

$$T_2 - T_1 = 0.83 \times 10^{-11} \times H^2.$$

For a field of 30,000 gauss this gives $T_2 - T_1 = 0.007°$, i.e. an effect that is quite negligible. However, equation (4) shows that at low temperatures the effect should be much greater; for not only is $T$ smaller, but $C_H$ is also smaller. Assuming for the moment that Curie's law were valid for solid oxygen, which is not the case, then at about 3° K., i.e. at $\frac{1}{100}$ of room temperature, $C_H$ might be expected to be about $\frac{1}{100}$ of its value for a gas at 293°. Then (4) would give

$$T_2 - T_1 \sim 0.8 . 10^{-7} H^2,$$

or, taking $H = 30,000$ gauss, $T_2 - T_1 \sim 70°$! This value is obviously absurd apart from the fact that equation (4) would then no longer be applicable, and signifies of course that the hypotheses made are not valid in this case; but it shows nevertheless that the magneto-caloric effect may possibly become large at low temperatures in the case of paramagnetic bodies.

In 1926 Debye and Giauque independently suggested that the magneto-caloric effect might be useful for obtaining low temperatures. Giauque himself began to prepare the necessary experiments. Similar work was commenced in the course of the next few years in nearly all the cryogenic laboratories of the world. Yet seven years elapsed before any very sensational results were obtained. This long period between an idea and its realisation requires a few words of explanation.

### III. iii. 2. *Problems connected with Magnetic Cooling*

The magnetic-cooling method has been developed with striking success in three laboratories: Berkeley, Leiden and Oxford, and has opened up new vistas for low-temperature physics. The men responsible for this development are Giauque and MacDougal, Wiersma and de Haas and Simon and Kürti. All have contributed valuable work to this new domain of very low temperatures.

The fundamental experimental arrangement for producing low temperatures with the magnetic method is clear enough, but in the details of the experiment plenty of room is left to the imagination of the individual experimenter. The properties of matter at extremely low temperatures differ so strongly from those obtaining even at the boiling point of helium that it took all low-temperature physicists a considerable time to become accustomed to the new conditions. The two essential properties which will concern us most here are specific heats and vapour pressures.

Until recently the lowest temperatures at which experiments could be carried out were of the order of magnitude of $1°$ K. Theoretically, it appeared possible with the help of the magnetic method to lower the temperature by at least one power of 10, possibly even more. Now at $0.1°$ K. the specific heat of all "ordinary" bodies—the meaning of the word "ordinary" will be explained shortly—is several powers of 10 smaller than at $1°$ K. According to Debye's law it should be one-thousandth of this value: in point of fact it is probably slightly greater. Moreover, not only are the vapour pressures of all bodies except helium to all intents and purposes zero, but the vapour pressure of helium itself is so small that nothing but a "complete" vacuum can exist at these temperatures, a vacuum much more complete than that which can be attained in any other way. These facts should naturally influence any experimental arrangement.

Now how about the specific heat of the so-called "extraordinary" bodies, which in our case are the paramagnetic substances with the help of which the low temperatures are to be obtained?

The degeneracy which gives rise to the magnetic entropy $R \ln P$ would, if it subsisted down to absolute zero, be in contradiction to Nernst's theorem. Moreover, we should expect that in a solid body the equivalent energy states corresponding to the degenerate term would not be completely identical but that interaction with neighbouring ions would cause the term to be slightly split up. Thus at sufficiently low temperatures the ions would no longer be evenly distributed over all the $P$ components of the term but would tend to be concentrated in the state with the lowest energy. As we approach absolute zero the magnetic entropy would therefore decrease. This would entail a decrease in the magneto-caloric effect and bring us to a natural limit below which the magnetic method of producing low temperatures would no longer be applicable. Simultaneously Curie's law for the susceptibility of the paramagnetic substance would cease to be valid.

Now the elimination of degeneracy and the concentration of the ions in the state of lowest energy must naturally lead to an anomaly in the specific heat. Under the somewhat inaccurate assumption that the $P$-fold term would split up into $P$ energetically equidistant components, Kürti showed that the anomalous specific heat must be very considerable as compared with the "ordinary" specific heat of the substance due to the thermal oscillations of the lattice, and more especially as compared with the specific heat of any other bodies present in the system. However, as no data existed as to the extent of the interaction, it was not known at what temperatures this effect was to be expected. Kürti therefore determined by direct experiment the specific heat in the helium region of gadolinium sulphate, a paramagnetic salt which was known to obey Curie's law down to very low temperatures (see chap. II, p. 232). The results are shown in fig. 15, p. 154. Down to about 6° K. the specific heat roughly follows Debye's law, but at lower temperatures a marked anomaly appears, the specific heat rising rapidly as the temperature falls. These results were later confirmed by Clark and Keesom and extended to still lower temperatures. It is interesting that at the lowest temperatures attained the susceptibility still obeys Curie's law, but Kürti showed that here only a very small deviation from the law was to be expected, as the specific heat is a much more sensitive criterion for the splitting of a term than the magnetic susceptibility. The lowest term of gadolinium sulphate is $^8S_{\frac{7}{2}}$ and thus eightfold degenerate, and Kürti assumed this term to be split into eight equidistant components. This, as Kürte himself states, is probably incorrect, as it may be shown that a twofold degeneracy should subsist down to very much lower temperatures. However, this would not alter the order of magnitude of Kürti's effect, and we may safely conclude that the magnetic method is limited for every substance by a temperature region in which the specific heat is abnormally great. This fact was used by Simon in his experimental arrangement, as will be shown in the next paragraph.

One of the fundamental difficulties now to be solved was how to measure the low temperatures which were to be attained by the magnetic method. Gas thermometry is not feasible below 1° K., as the vapour pressures of helium become too small and the Knudsen effect makes accurate measurements impossible. It is perfectly natural that gases should fail as thermometer substances at the same temperatures at which they cease to be useful for lowering the temperature. Where inner atomic processes are necessary to produce refrigeration they are equally necessary to determine the refrigeration they produce. Thus it is quite logical that the magnetic susceptibilities of the paramagnetic salts furnishing the magneto-caloric effect should be used to measure the low temperatures thus attained. As we have just seen, it is profitable to work with substances that follow Curie's law down to very low temperatures. As the example of gadolinium sulphate showed that Curie's law is observed even at temperatures at which the specific heat is already markedly anomalous, we may fairly safely assume that this law is approximately valid down to the lowest temperatures attainable with the magnetic method. This assumption has been made by all authors working in this field, although in some cases plausible corrections have been applied. In all cases upper limiting temperatures were thus deduced, and we may assume that the low temperatures reached were certainly not higher than those published. However, we must note that these temperature determinations do not possess the accuracy of absolute measurements.

The link between the thermodynamic and the magnetic or any other temperature scale may be effected with the help of calorimetric measurements according to the Second Law, as shown in Part I, chap. IV. Though these measurements have not yet all been carried out, or at any rate published, the method is perfectly clear and there is no reason to doubt that we shall soon be able to measure temperatures of $0.01°$ K. just as accurately as we can now determine temperatures a thousand times as high.

Consider a $T$-$S$ diagram of the paramagnetic substance with the help of which the low temperature is to be attained. The curve in fig. 1 shows temperature as a function of entropy for $H = 0$. Let us suppose that the temperature $T_0$ is high enough

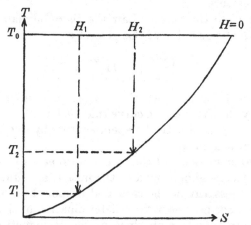

Fig. 1. Absolute determination of very low temperatures.

to be measured by a gas thermometer and at the same time low enough for the entropy due to the ordinary vibrations of the crystal lattice to be neglected. Such a temperature may always be found, but in fact the second condition is not essential and merely assumed for the sake of simplicity. The entropy at $T_0$ and $H = 0$ will then be $R \ln P$, as we have just seen. At finite field strengths we shall have

$$S_{T_0 H} = S_{T_0 H = 0} + \int_0^H \left( \frac{\partial \sigma}{\partial T} \right)_{H = 0} dH,$$

and the relation between $\sigma$, $T$ and $H$ in the neighbourhood of $T_0$ may be determined by direct experiment.

At lower temperatures we may expect the entropy for $H = 0$

to fall off as shown in the figure as a result of the splitting of the $P$-fold degenerate term. We may determine the shape of the curve by a series of adiabatic demagnetisations along the vertical lines descending from various values of the field strength. The temperatures reached by these demagnetisations will then be determined with some arbitrary thermometer, say from the magnetic susceptibility with the help of Curie's law. Temperatures thus determined may be indicated by an asterisk. Now in this same temperature scale we may by direct calorimetric measurement determine the specific heat at the low temperature.

According to the Second Law the thermodynamical temperature $T$ at this point will be given by

$$T = \left(\frac{\partial U}{\partial T^*}\right)_{H=0} \Big/ \left(\frac{\partial S}{\partial T^*}\right)_{H=0}.$$

$(\partial U/\partial T^*)$ is the specific heat that we have measured, $(\partial S/\partial T^*)$ is given by the shape of the curve that we have found. We are thus able to determine the true temperature by a combination of direct measurements.

Here we have assumed that $\sigma$ is known as a function of $T$ and $H$ in the neighbourhood of the initial temperature. This need not necessarily be the case, and Keesom has shown that even this is not necessary for a determination of the thermodynamical temperature obtained by the demagnetisation. In fig. 2 two curves of constant field strength $H_1$ and $H_2$ have been included apart from $H = 0$. Here we shall not assume that any of these curves are known, but merely that $T_1$ may be determined with a gas thermometer. Now suppose adiabatic demagnetisations to be carried out from the points 1 and 2 on the figure to $H = 0$. The resulting points on the curve for $H = 0$ will be 3 and 4. We may now measure the quantity of heat $Q$ needed to bring us directly from 4 to 3 in the absence of a field. This merely necessitates the use of an arbitrary thermometer that will always give the same reading at point 3. Finally, suppose an adiabatic demagnetisation to be effected from $H_2$ to $H_1$, i.e. from point 2 to point 5. We may then determine the quantity of heat $Q_1$ needed to bring us from 5 to 1 at the field strength $H_1$.

Now according to the Second Law $Q = \int T dS$, so that the $Q$'s appear as areas in fig. 2. The ratio $T/T_1$ is equal to the ratio of the areas of the rectangles 3687 and 1287. If $H_1$ and $H_2$ are sufficiently close together this ratio will be equal to that of the areas 3487 and 1587, which are equal to $Q$ and $Q_1$ respectively. Therefore if the true thermodynamic temperature at 3 be $T$, $T/T_1 = Q/Q_1$. Thus, as $T_1$, $Q_1$ and $Q$ are known, $T$ may be computed. We may therefore determine the thermodynamic temperature from calorimetric measurements alone without knowing the $\sigma$-$T$-$H$ relations even at higher temperatures.

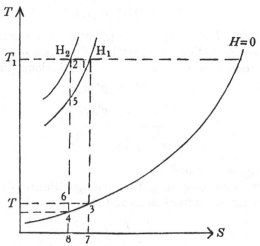

Fig. 2. Absolute determination of very low temperatures.

Recently de Haas and Wiersma showed that the absolute temperature scale may be determined at very low temperatures from magnetic measurements alone. The authors found that in caesiumtitanic alum $(Cs_2So_4Ti_2(SO_4)_3 . 24H_2O)$ adiabatic demagnetisation from various high field strengths to about 100 gauss produces no change in the magnetisation $\sigma$, that is to say, under these conditions a process in which $dS = 0$ is simultaneously characterised by $d\sigma = 0$. Since

$$dS = \left\{ \frac{1}{T} \left( \frac{\partial U}{\partial \sigma} \right)_T - \frac{H}{T} \right\} d\sigma + \left( \frac{\partial U}{\partial T} \right)_\sigma \frac{1}{T} dT,$$

$dS = d\sigma = 0$ gives $\left(\dfrac{\partial U}{\partial T}\right)_\sigma = 0$ for all "high" values of $\sigma$. Since, according to the Second Law,

$$\frac{\partial}{\partial T}\left\{\frac{1}{T}\left(\frac{\partial U}{\partial \sigma}\right)_T - \frac{H}{T}\right\} = \frac{\partial}{\partial \sigma}\left\{\frac{1}{T}\left(\frac{\partial U}{\partial T}\right)_\sigma\right\},$$

we have
$$\frac{1}{T}\left(\frac{\partial U}{\partial \sigma}\right) - \frac{H}{T} = -f(\sigma)$$

and at the same time
$$\left(\frac{\partial U}{\partial \sigma}\right)_T = g'(\sigma).$$

Thus
$$H = T \cdot f(\sigma) + g'(\sigma).$$

Now if $H_1$, $T_1$ and $H_2$, $T_2$ are points on two known saturation curves at helium temperatures for which $\sigma$ has a value $\sigma_0$ which remains constant during adiabatic demagnetisation, then
$$H_1 = T_1 \cdot f(\sigma_0) + g'(\sigma_0),$$
$$H_2 = T_2 \cdot f(\sigma_0) + g'(\sigma_0)$$

and
$$g'(\sigma_0) = \frac{T_1 T_2}{T_2 - T_1}\left(\frac{H_1}{T_1} - \frac{H_2}{T_2}\right),$$

and is thus determined.

As $\sigma$ remains constant during demagnetisation $f$ and $g'$ are constants for one experiment, and thus

$$\frac{H - g'}{T} = f.$$

Thus if the field strengths and temperatures at the beginning and end of a demagnetisation experiment are $H_b$, $T_b$ and $H_e$, $T_e$ respectively,

$$\frac{H_b - g'}{T_b} = \frac{H_e - g'}{T_e}; \quad \text{i.e.} \quad \frac{T_e}{T_b} = \frac{H_e - g'}{H_b - g'}.$$

As the other three quantities are known, $T_e$ may be determined.

Another point which has received considerable attention is the choice of a suitable paramagnetic substance. It is obviously important to select a substance which follows the Curie-

Langevin law down to the lowest temperatures attainable with liquid helium, i.e. with comparatively "free magnetic dipoles". For if the effects of interaction are noticeable at these temperatures they will be sure to disturb the process lower down. Early experiments of Kamerlingh Onnes had shown that, as was to be expected, the most ideal paramagnetic bodies in this sense are dilute salts, i.e. compounds in which the individual paramagnetic ions are situated far from one another and each surrounded by diamagnetic particles. The classical example of this type of salt is gadolinium sulphate octohydrate $Gd_2(SO_4).8H_2O$, the susceptibility of which was studied in detail by Kamerlingh Onnes and Woltjer and found to follow the corrected Langevin formula as far down as $1\cdot2°$ K. This salt was therefore suggested by Debye and Giauque as probably the most favourable example, and the latter took great trouble to collect large quantities of gadolinium sulphate for his experiments. After Kürti had found that the specific heat of gadolinium sulphate becomes anomalous at comparatively high temperatures, the Leiden physicists began experimenting on a number of other salts and found potassium chromic alum to give the best results.

The most systematic researches in this direction are due to Kürti and Simon, who attempted with considerable success to classify the various substances by means of a characteristic temperature $\theta = U/k$, where $k$ is Boltzmann's constant and $U$ the energy difference between consecutive states of a degenerate term split up by interaction. These states are here again assumed as a first approximation to be equidistant. Simon and Kürti showed how $\theta$ may be determined for a given substance at temperatures easily obtainable with liquid helium. The anomaly in the specific heat will have a maximum at $T \sim \theta$, and here too the deviations from Curie's law will be considerable. Thus the lowest temperatures to which a given paramagnetic body will bring us is of the order of magnitude of $\theta$. The smaller $\theta$, the more suitable the substance for producing low temperatures.

Now if the specific heat of a substance is very small, it will be difficult to maintain it at a given low temperature; the substance will warm up rapidly after the lowest temperature is reached. Thus in order to carry out an experiment at a given low temperature, it will be useful to employ a paramagnetic substance the $\theta$ of which is roughly equal to this temperature; for here the specific heat is greatest and the temperature will remain practically constant for a long period of time.

### III. III. 3. *Experimental Procedure and Results*

In the first Leiden experiments the paramagnetic substance was suspended from a magnetic balance (see chap. II, § 2) in a capsule within a closed glass tube mounted in a helium cryostat between the poles of a large electromagnet ($H \sim 30,000$ gauss). The magnetic balance was designed to measure the magnetic susceptibility of the salt and thus to determine the temperature. The capsule and tube were filled with helium at a very low pressure, which effected thermal contact during the cooling of the tube in the cryostat and eliminated the heat effect produced on turning on the magnet. When the magnetic field was switched off, the substance was cooled down to such a low temperature that the traces of helium were condensed and the vapour pressure sank to an excellent vacuum, thus automatically isolating the capsule.

There were several drawbacks to this apparatus. The magnetic balance is very sensitive and not simple to operate, the cooling of the sample to helium temperatures took several hours and the removal of the magneto-caloric heat generated by applying the magnetic field was slow. It was therefore necessary to keep the field operating for a considerable time. Moreover, the substance warmed up rather quickly after demagnetisation, especially at the lowest temperatures, a fact which the authors ascribe to the condensation of helium. Another disadvantage was the impossibility of decreasing the field of the large electromagnet rapidly to less than a few hundred gauss. Nevertheless, very satisfactory results were obtained with this apparatus, more especially in the later experiments. With potassium chromic alum a record temperature of $0 \cdot 05°$ K. was reached, as determined from the susceptibility measurements.

Shortly afterwards a second apparatus was built, in which

a number of disadvantages were eliminated. The magnetic balance was replaced by an induction arrangement, and the vessel containing the paramagnetic substance could be evacuated by a pump after the field had been applied. Moreover, the container was larger than in the previous arrangement. The reduction of the magnetic field to zero was accomplished by swivelling the entire cryostat out of the interferrum, after which a solenoid was placed around it to measure the susceptibility.

This apparatus was necessarily rather complicated, as the movable cryostat had to be connected to the pump by a ground joint and the pump lead was long. It has, however, yielded the lowest temperatures yet obtained. With a solution of $K_2SO_4Cr_2(SO_4)_3.24H_2O$ and 14·4 parts of diamagnetic $K_2SO_4Al_2(SO_4)_3.24H_2O$ Wiersma and de Haas reached a temperature of $0·0044°$ K. This is probably a much lower temperature than was originally expected by the first advocates of the magnetic method. The perspective it offers to future research is literally boundless and the work of Wiersma and de Haas will remain as one of the classical masterpieces of low-temperature physics.

The main feature of Giauque's apparatus is the employment of a solenoid without iron as a magnet. This entails several obvious advantages, especially the possibility of cooling very large volumes, but the field strengths reached are naturally smaller than those produced with the Leiden magnet. Moreover, Giauque has hitherto published only results obtained with gadolinium sulphate, which is now known to be an unsatisfactory substance. Thus the temperatures attained are considerably higher than those reached at Leiden. However, a number of valuable determinations were made of the thermodynamical data of gadolinium sulphate.

It is the work of Simon and Kürti that shows the greatest mastery of the new technique. These authors possess no full-sized helium liquefier and the liquid helium needed for the experiments is produced in small quantities inside the apparatus, amounting in all to a few cubic centimetres. This is effected with the help of a miniature liquefier, as described in Part I, p. 51. It necessitates work on a small scale, and Simon and Kürti have turned this necessity into an advantage. In what they would probably term their large-size apparatus (fig. 3) the paramagnetic salt is cooled to about 1° K. by some 2 c.c. of liquid helium in the small vessel $C$ specially isolated from the rest of the apparatus. The container $B$ with the salt is attached to this vessel with a ground joint by means of tap-grease, which was found to hold against the small differences of pressure involved. The lowest temperature in the liquid helium is reached an hour after the commencement of the experiment. Thereupon the contact helium is evacuated through the tube $R$, the magnetic field being applied when a pressure of about $10^{-2}$ mm. is reached. After about 10 minutes, when the pressure has fallen to $10^{-4}$ mm., the field is switched off and an excellent vacuum automatically produced by condensation of the remnants of helium. The susceptibility is measured by an inductance method to determine the temperature. The inductance coils are shown at $Y$ on fig. 3.

However, as the authors have no large magnet at their disposal, they have to overcharge their magnet to obtain the necessary high fields. The magnetocaloric warming-up period must therefore be reduced as far as possible. A second apparatus was therefore built, in which the field operates for a still shorter period. Though the arrangement has certain disadvantages it is so simple that it evidently points the way for new technical development.

The substance is packed loosely in a closed thin-walled glass tube containing 1 cm. of helium at room temperature and sealed off. The capsule is placed directly in liquid helium, the temperature of which is reduced as far as possible. Equilibrium with the bath is reached in a few minutes. When the magnet is turned on the heat of magnetisation is removed in 3 minutes. For although the helium pressure is very small, the heating effect causes gas to be desorbed from the substance, which promptly carries the heat of magnetisation to the walls of the capsule. After demagnetisation a perfect vacuum is immediately attained. Practically no thermal contact exists between the loosely packed coarse powder and the walls of the container, so that the sample warms up at a rate of 0·0003° per minute. Under these conditions experiments can be carried

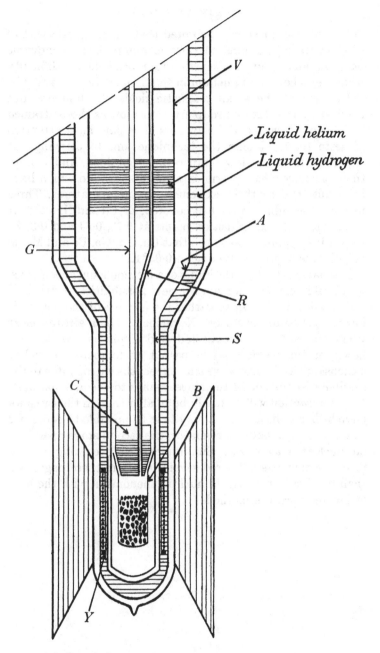

Fig. 3. Simon's demagnetisation apparatus.

out with ease. It should be noted that the magnetic part of this apparatus contains no vacuum pumps and no movable parts of any kind. Its construction presents no difficulty whatever where liquid helium in small quantities is available.

In view of the small magnetic fields the temperatures reached by the Oxford group are not as low as those attained at Leiden. However, these authors have paid most attention hitherto to developing the technique and to carrying out experiments at the low temperatures obtained, and it is to them that the first experimental results are due which have been obtained at these extremely low temperatures. Three new supra-conductors were discovered, Zr, Cd and Hf (cf. Part IV, chap. II), with transition points at $0.7°$, $0.54$ and $0.3°$ K. respectively, and it was found that Cu, Au, Ge, Be and Mg do not become supra-conducting at $0.05°$ K.

It is natural that, as the limits of a method loom ahead of us, we should scan the horizon for new possibilities. Until now we have hardly begun to gather the harvest of the magnetic low-temperature generator. Nevertheless, it is already clear where the end is to be expected, and Simon has pointed out how even this barrier may be passed. At temperatures below a thousandth of a degree we may expect the entropy due to the random distribution of the nuclear spins to become manifest. Here interaction will certainly be smaller than in the cases we have hitherto discussed. It may thus be possible to construct a two-stage magnetic refrigerator, the second stage operating on nuclear magnetism. Simultaneously, according to the principle that has appeared several times in these pages, we shall be able to investigate nuclear phenomena with the help of low-temperature methods.

# Part IV

## THE "FREE" ELECTRON

### CHAPTER I

### CONDUCTIVITY AT LOW TEMPERATURES

#### IV. I. 1. *Introductory*

To form a picture of the magnetic properties of matter it
appeared advantageous to stress such electronic properties as
become conspicuous when the electron is considered as a
constituent part of an atom or molecule. In these chapters, in
which we shall attempt to show what light low-temperature
research has thrown on the passage of heat and electricity
through solids, these same electrons appear more or less as
carriers of heat and electricity. Though it is an open question
how far the so-called conductivity electrons are really free,
the term "free" electron is employed in order to stress the
carrier properties of the electrons and to imply that their
specific nature as constituent parts plays a less important rôle
in these processes.

However, though comparative unanimity reigns as to the
fact that the thermal and electrical conductivity of metals is
to be ascribed to the motion of electrons, and though all
existing theories have to some extent been corroborated by
experiment, yet none of these theories can be said to have
solved the problem. A number of experimental results are in
flat contradiction to current theory, such as, for instance, the
temperature dependence of thermal and electrical resistances
at very low temperatures.

It is evident that the primitive pictures that helped us to
follow the processes considered in the first parts of this book

are insufficient to describe conductivity phenomena. This is natural and merely means that the farther we proceed from the macroscopic occurrences that may be treated with the help of classical thermodynamics and mechanics to processes occurring in electronic dimensions, the less our every-day pictures are applicable and the more quantum mechanical concepts move into the foreground. In thermal and electrical conductivities we have to do not merely with a stream of electrons moving through matter according to the classical concepts of Lorenz or according to the "corrected" concepts of Sommerfeld, but with electronic waves interacting with the waves of the periodical lattice, and unfortunately the formulae deduced from these concepts are still to some extent at variance with our observations. In this domain of physics theory has not yet outstripped experiment, and up till now a concise survey of experimental facts appears to give us a clearer conception of this sphere of nature than any pictorial illustration.

In his book, *The Metallic State*, W. Hume Rothery has given an excellent résumé of classical and current theories of conductivity. We see no reason to expatiate on or repeat his critical discussion, but would rather recommend it to the reader as a whole, leaving it to the author to bring it up to date in a later edition. In the following chapters we shall on the whole keep to the bare experimental facts, and the word electron will hardly appear. Moreover, since conductivity has already received considerable attention in literature, we may confine ourselves to a very brief sketch of the historical facts and to a discussion of the latest developments of low-temperature research in this field.

### IV. 1. 2. *Thermal Conductivity*

The thermal conductivity of solids at low temperatures has only quite recently become an object of concentrated research. A number of reasons may be given for the comparative lack of interest hitherto shown. In the first place heat conductivity was formerly not classed as a primary physical property of matter but rather as a subsidiary characteristic of more technical interest. Moreover, accurate measurements of heat conductivity, especially at low temperatures, are far more difficult than electrical resistance measurements. It is far easier to maintain an electrical potential difference at the ends of a rod or wire than a difference of temperature, and whereas electrical conductivity can be investigated at a definite uniform temperature, thermal conductivity cannot. In none of the typical forms of apparatus designed for low-temperature work had it proved necessary to keep two ends of a specimen at well-defined different temperatures. But the chief reason that heat resistance was not studied in detail until 1930 lay in the fact that no very interesting discoveries were expected in this field. It was known long since that at room temperature the heat conductivity of metals was almost independent of temperature, rising slightly as the temperature fell, and that at low temperatures the dependence on temperature grew gradually more pronounced. According to existing theory a quadratic relation between thermal resistance and temperature was to be expected in the neighbourhood of absolute zero.

In point of fact, when a systematic investigation of thermal resistance was begun at Leiden, nothing very striking was at first discovered. Around liquid-air temperatures the heat resistance of tin fell steadily with an increasing gradient, and at the temperatures of liquid hydrogen the curve showed a tendency to straighten out. However, it appeared unlikely that an extrapolation to $T = 0$ would lead to zero resistance. It was not till the experiments were continued at the temperatures of liquid helium that really startling results were obtained. Somewhere between hydrogen and helium temperatures, probably at about 9° K., the curves giving the relation between thermal resistance and temperature, in the case of lead and tin, pass through a very pronounced minimum. Below this minimum the thermal resistance increases rapidly as the temperature is lowered, and in the case of lead the resistance at 2·2° K. is equal to that at 50° K. and practically as great as at the boiling point of oxygen. Fig. 1 is reproduced from the original paper of de Haas and Bremmer. Very similar curves were obtained for tin and indium. In the case of the alloy $PbTl_2$ the thermal resistance increases slightly with falling temperature throughout the temperature scale, so that no actual minimum occurs; but the sharp increase in the region of liquid helium is the same as for pure metals.

These facts are clearly opposed to current theory, and the authors are disposed to consider that possibly secondary effects, such as impurities or non-homogeneities, may play an

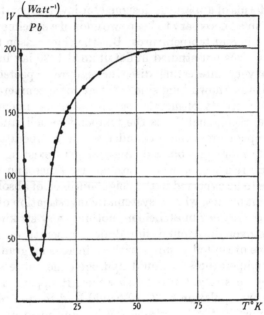

Fig. 1. Heat resistance of lead.

important part. Nevertheless, it is clear that the sharp increase of thermal resistance below the boiling point of helium will be of great influence in the designing of apparatus for low-temperature work.

The increase of thermal resistance at low temperatures is not confined to metals and their alloys. The resistance of crystallised quartz parallel to the principal axis shows the same minimum around 10° K. as was demonstrated by recent experiments of de Haas and Biermasz, so that an increase of thermal resistance in the helium region appears to be a universal property of matter.

A number of appliances have been developed for measuring heat conductivity at low temperatures. As a rule the specimen has the form of a rod, one end of which is in thermal contact with the liquid in the cryostat, while the other is kept at a constant and slightly higher temperature by an electric heater. Roughly speaking the measurement consists in determining the quantity of heat which must be applied to one end of the rod in order to maintain this end at a certain temperature, the heat input being equivalent to the quantity of heat transported along the rod. Naturally a number of corrections must be applied, which we cannot here discuss. The temperature difference between the ends of the rod was at first measured with a thermocouple. However, at very low temperatures the thermoelectric forces are too small for accurate determinations, especially when small differences of temperature are to be measured. De Haas and his collaborators therefore resorted at first to gas thermometers and later to vapour-pressure thermometers as the most accurate temperature gauge in the helium region.

Fig. 2 shows a sketch of the apparatus used by de Haas and Bremmer in 1931. $V$ is a brass vessel that can be evacuated through $t$, and which is placed in the cryostat. The lower end of the specimen $W$ is in contact with the liquid outside $V$. The upper end is heated with the help of coil $K$ and the temperature measured with a gas thermometer, the bulb $R$ of which is in contact with the top of the rod. A German silver spiral capillary connects the bulb with the other parts of the gas thermometer.

In 1934 de Haas and Capel developed a new form of apparatus. Here two rods $W_1$ and $W_2$ (fig. 3) serve as specimens and are connected below by a copper rod $W$ bent to a U. The upper ends of the specimens are joined to two glass tubes $u_1$ and $u_2$, which serve as reservoirs for two vapour-pressure cryostats filled with hydrogen or helium. One of these tubes contains the heating coil $k$, the other is connected to a vacuum pump, which maintains a constant vapour pressure on this side and thus a constant temperature. A definite current in the heating coil thus keeps up a definite temperature at the other side of the apparatus, which is constant under stationary conditions and can be measured by the vapour pressure. This apparatus is mounted in a glass vacuum jacket $B$ sealed through the top of the cryostat. Fig. 3 shows merely a schematic sketch of part of the apparatus, which is in fact rather more complicated.

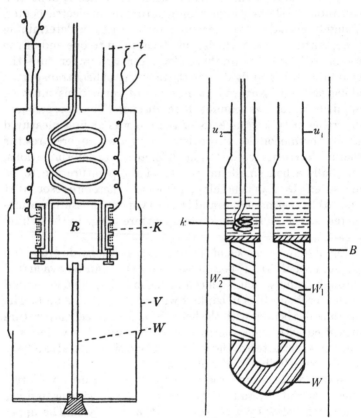

Fig. 2. Apparatus for measuring heat conductivity.

Fig. 3. Apparatus for measuring heat conductivity.

## IV. 1. 3. *Electrical Conductivity*

The electrical resistance of pure metals decreases as the temperature is lowered, at first more and more rapidly, then more slowly, finally approaching a value independent of the temperature, which is known as the residual resistance and is the smaller the purer the specimen. It appears highly probable that at absolute zero the resistance of an absolutely pure metal single crystal would be equal to zero. Fig. 4 shows the resistance of a number of gold wires of different purity as a function of temperature. The purer the gold, the lower the curve. Moreover, it is clear from the figure that the curves may be made to coincide by shifting them parallel to the resistance

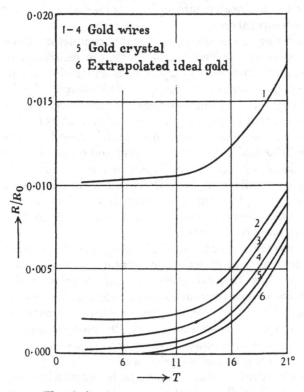

Fig. 4. Electrical resistance of gold wires of different purity.

axis. This is known as Matthiessen's rule and signifies that the residual resistance due to impurities and possibly non-homogeneities of the material is independent of the tempera-ture. For not too impure material this rule is accurately valid at liquid-helium temperatures and even at the temperatures of liquid hydrogen it gives a very good approximation. It is thus possible to determine the "ideal" resistance of an absolutely pure metal by displacing the resistance of a very pure specimen in such a way that the curve passes through the origin. The impurities capable of producing a considerable residuary resistance are very small; in fact the electrical resistance is one of the most sensitive methods of testing the purity of a metal. Metals which by any other test appear perfectly pure are frequently not nearly pure enough for resistance thermometry at low temperatures.

According to the theory of metallic conductors, developed some years ago by Peierls and Bloch, the resistance at very low temperatures should vary as $T^5$. A critical discussion of the whole experimental material up to 1932, compiled by de Haas and Voogt, showed that whereas the resistance of most pure metals seemed to vary as some power of $T$ at very low temperatures, the power 5 was never quite reached. In the case of the "classical" metals, gold, silver and copper powers of $T$ were found between 4·2 and 4·5, whereas a number of other metals gave still lower powers down to 2·8. Recent experiments on very pure platinum have shown that in fact no power of $T$ gives the resistance-temperature relation accurately, which is really more complicated.

The electrical resistance of alloys is generally much greater than that of pure metals. More striking than the resistance itself are the thermal coefficients of electrical resistance, which are always very much smaller than in the case of pure metals. Not only in such well-known alloys as constantan and manganin is the resistance practically independent of temperature, but in almost all alloys the resistance at helium temperatures is almost as great as at room temperature.

The electrical resistance of a metal increases when placed in a magnetic field. This effect, which is most apparent when the field is perpendicular to the electric current, is small in the case of most metals and could be studied only in very strong fields.

The work of Kapitza on the production of magnetic fields up to 300 kilogauss and of 0·01 sec. duration and of the measurement of electrical resistance under these conditions is one of the most brilliant feats of modern experimental technique. Unfortunately we cannot here describe these experiments, which are connected with our subject only in so far as the influence of a magnetic field on electrical resistance increases as the temperature is lowered. As a rule the relation between magnetic field strength and electrical resistance is given by a curve of the type shown in fig. 5. In small fields the relation is quadratic, in large fields linear, the region of transition from one type to the other being different for every metal. According to Kapitza the linear relation is the "true"

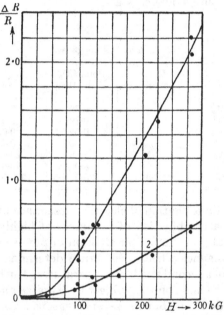

Fig. 5. Electrical resistance of beryllium in a magnetic field. Magnetic field perpendicular to current. (1) Liquid hydrogen temperature, (2) room temperature.

effect, the quadratic curve being due to the same disturbances and impurities that give rise to the residual resistance. The curves are very sensitive to impurities and to mechanical treatment of the specimens.

To give a survey of the order of magnitude involved a few of Kapitza's results are listed in Table XXV.

From this table it is evident first that the effect increases considerably with falling temperature, and secondly that in the doubtful metals As, Sb and Bi the electrical resistance is far more sensitive to magnetic fields than in any of the others. In this group bismuth plays a particularly prominent part. Moreover, experiments with single crystals of bismuth showed that the change of resistance in a magnetic field, especially at low temperatures, is highly dependent on the orientation of the crystal. This, combined with the sensitivity to impurities and mechanical strain, necessitated the preparation of exceptionally pure single crystals.

### Table XXV

$\Delta R/R$ at 300 kilogauss.

| Metal | Room temperature | Solid $CO_2$ with ether | Liquid nitrogen |
|---|---|---|---|
| Mg | 0·17 | 0·32 | 2·84 |
| Cd | 0·08 | 0·19 | 0·93 |
| Hg | . | 0·02 | 0·05 |
| Al | . | 0·09 | 0·65 |
| As | 1·1 | 2·85 | 29·0 |
| Sb | 3·5 | 8·0 | 40·0 |
| Bi | 37·0 | 196·0 | 1360·0 |

Thus the investigation of electrical resistance in a magnetic field led to a combination of a number of highly specialised experimental procedures—production of pure single crystals, very high magnetic fields, resistance determinations in one hundredth of a second and low temperatures. Much interesting work has been done on this subject by Kapitza and by Schubnikow and de Haas, but for several reasons a complete combination of all the above-mentioned points has not yet been achieved.

Bismuth crystallises in a rhombohedric face-centred lattice and thus has three binary axes perpendicular to the principal trigonal axis. In the experiments of Schubnikow and de Haas a single crystal rod of bismuth was used with the principal axis in the axis of the rod. The crystals were exceptionally pure. In all experiments the magnetic field was perpendicular to the main axis, i.e. to the axis of the rod. The change of resistance in a magnetic field was measured at the temperatures of liquid air and of liquid hydrogen in various directions of the field with respect to the binary axis of the crystal. The influence of the field on the resistance increased very rapidly as the temperature was lowered, so that we may confine ourselves to the results obtained at hydrogen temperatures.

Fig. 6 shows $R_H/R_0$ at $T = 14 \cdot 15°$ at various field strengths up to 30 kilogauss. $R_H$ is the resistance in the field $H$ at $14 \cdot 15°$ K., $R_0$ the resistance without a field at $0°$ C. The upper curve was obtained with $H \parallel [11\bar{2}]$, the lower with $H \perp [11\bar{2}]$. These curves

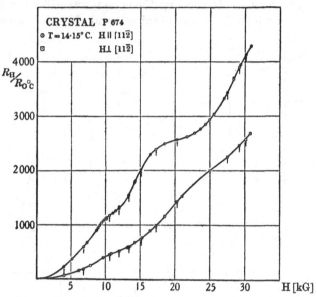

Fig. 6. Electrical resistance of bismuth in a magnetic field.

give the maximum and minimum values of $R_H/R_0$. Clearly the simple relation found by Kapitza between $H$ and $R$ does not hold for single crystals of bismuth at hydrogen temperatures. More especially the upper curve shows several humps, which cannot be explained as due to impurities. At 30,910 gauss $R_H/R_0$ attains the value of 4295. At this temperature $R/R_0$, where $R$ is the resistance at $14 \cdot 15°$ K. without a magnetic field, is equal to $0 \cdot 0252$; the actual change of resistance due to the magnetic field is thus

$$R_H/R = 176,000.$$

As the crystal is turned in the magnetic field through $360°$, a sixfold periodicity of $R_H/R_0$ is observed. In weak fields this leads to sin-curves with a period of 60°. But as the field is increased these curves begin to degenerate, subsidiary minima and maxima appear and the higher the fields and the lower the temperature, the more extravagant do the curves become. Fig. 7, which is reproduced from a paper of Schubnikow and de Haas, shows a set of curves obtained at 14·15° K. in fields up to 30 kilogauss. In this case the crystal was still purer than that with which fig. 6 was obtained, and thus the values of

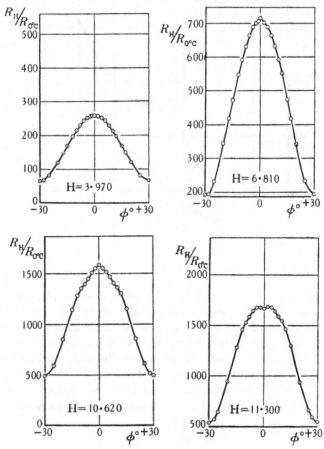

Fig. 7. Electrical resistance of bismuth in a magnetic field.
Rotation diagrams round the hexagonal axis.

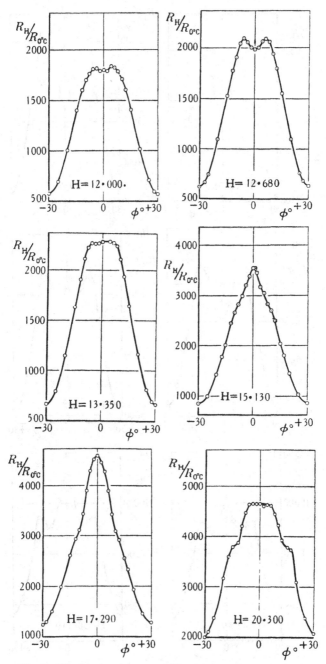

Fig. 7. Electrical resistance of bismuth in a magnetic field.
Rotation diagrams round the hexagonal axis.

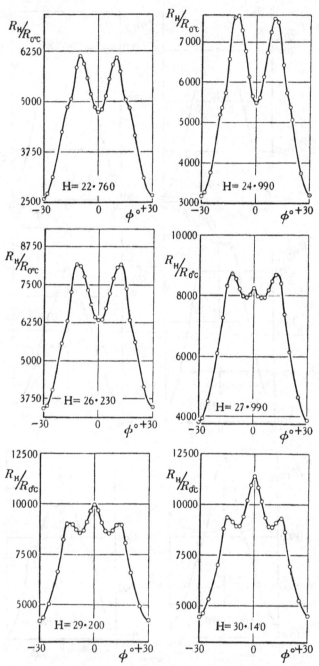

Fig. 7. Electrical resistance of bismuth in a magnetic field.
Rotation diagrams around the hexagonal axis.

$R/R_0$ were almost three times as great as those mentioned above.

Bismuth is not the only substance that exhibits these strange properties. Later experiments of de Haas and Blom on gallium, which is tetragonal, show a similar dependence of the resistance on the direction of a magnetic field, though the absolute change of resistance is very much smaller than in the case of bismuth.

The relation between electrical resistance and magnetic field strength is evidently connected with the diamagnetic properties of metals. According to experiments of de Haas and

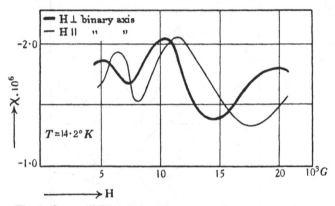

Fig. 8. Susceptibility of bismuth as a function of field strength.

van Alphen, the resistance of all metals with high diamagnetic susceptibility is particularly sensitive to magnetic fields. These authors measured the susceptibility of bismuth crystals as a function of the field strength and of the direction of the field with respect to the axes of the crystal. Whereas in fields parallel to the principal trigonal axis the susceptibility is independent of the field strength at all temperatures, in transversal fields this is no longer the case in the hydrogen region. Fig. 8 shows that at 14·2° K. the susceptibility is a periodic function of the field strength and that the maxima and minima lie at different values of H according as the field is parallel or perpendicular to one of the binary axes. In fig. 9 the magnetisation is shown at various field strengths as a function of the angle between the field and a binary axis. The correspondence with the resistance curves of fig. 6 is obvious.

In this connection we may refer to a paper of Ehrenfest, in which the abnormally high diamagnetism of metals such as

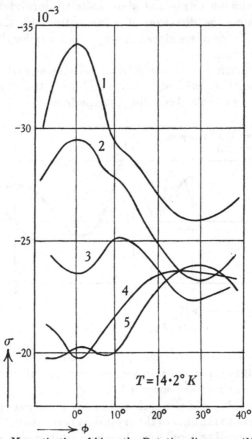

Fig. 9. Magnetisation of bismuth. Rotation diagrams. (1) 18·8, (2) 17·4, (3) 15·7, (4) 13·7, (5) 11·9 kilogauss.

bismuth, which decreases abruptly on melting, is ascribed to orbits of electron groups which have fixed positions in the crystal lattice. The question of how many separate directions in a bismuth crystal must be endowed with independent values of electrical resistance and diamagnetic susceptibility has been discussed theoretically by Stierstadt and Kohler.

# CHAPTER II

## SUPRA-CONDUCTIVITY

### IV. II. 1. *Introductory*

At low temperatures, mostly attainable only with liquid helium, a number of metals show supra-conductivity. Within a very narrow range of temperature, which is usually less than a thousandth of a degree, and in the case of single crystals probably a "geometric point" of temperature, the electrical resistance suddenly drops to zero, i.e. to a value so small that it could not as yet be determined.

Until recently only comparatively few supra-conductors were known, but at every new step towards absolute zero some fresh examples are discovered. However, it is by no means clear whether at sufficiently low temperatures all metals become supra-conducting. No clear relation seems to exist between the transition point, the temperature at which a metal loses its resistance, and any other physical property, nor are supra-conductors confined to certain groups of the periodic system.

In Table XXVI a list is given of all supra-conductors hitherto known among pure metals with their transition points:

### Table XXVI. *Supra-conducting metals*

|  | ° K. |  | ° K. |
|---|---|---|---|
| Nb | 8·4 | Ti | 1·77 |
| Pb | 7·26 | Th | 1·43 |
| La | 4·71 | Al | 1·14 |
| Ta | 4·38 | Ga | 1·10 |
| Hg | 4·12 | Zn | 0·79 |
| Sn (white) | 3·69 | Zr | 0·70 |
| In | 3·37 | Cd | 0·60 |
| Tl | 2·38 | Hf | 0·3–0·4 |

Apart from pure metals a large number of alloys become supra-conducting, all alloys containing one or more supra-conducting metals and several alloys neither of whose components becomes supra-conducting by itself. The classical example for this type is the alloy gold-bismuth. For the relations existing between the transition point and the constitution diagram of an alloy we must refer to the original papers of Meissner and Allen.

Supra-conductivity has also been observed in intermetallic compounds in the strict sense of the word and in a number of nitrides and carbides. In these cases the transition point is usually spread out along an interval of temperature, which may be several degrees wide.

Supra-conductivity is the kind of phenomenon that every physicist would like to have discovered; it is very striking and easy to verify. Apart from liquid helium practically no apparatus is required. It is natural that countless experiments should have been made to study supra-conductivity. Only comparatively few of these experiments are of interest from the point of view of experimental technique, and not very many have led to tangible results. Almost all attempts at an *experimentum crucis* to decide some point of discussion on supra-conductivity have merely complicated matters more by showing that both views were false. Evidently supra-conductivity is a very subtle phenomenon and it is surprising but true that we do not know much more about it to-day than we did twenty years ago.

## IV. II. 2. *Permanent Currents*

When at a temperature below its transition point a closed supra-conducting circuit is placed in a magnetic field, a current is induced which, owing to the absence of electrical resistance, continues to flow for an indefinite period. A similar current in the opposite sense is obtained if the supra-conducting circuit is cooled below its transition point in a magnetic field and the field then shut off. Supra-conductors, in which a current had been induced at Leiden, have been transported to Amsterdam, where it was shown that the current still flowed. In all experiments these "permanent currents" continued without any perceptible decrease in strength until the liquid helium evaporated and the temperature rose above the transition point. Permanent currents may easily be detected by means of a compass needle, which is deflected when brought into the vicinity of the supra-conductor.

A potential applied to two parallel circuits produces a current distribution inversely proportional to the resistance of the circuits. How is the current distributed in two parallel supra-conducting circuits, in both of which the resistance is zero? This was the question answered by some very attractive experiments of Sizoo at Leiden. Two rectangular circuits of tin wire were connected in parallel. The wires were mounted in a plane in such a way that a potential applied to the leads produced two currents in opposite directions. One of the wires was twice as thick as the other. A compass needle mounted in the plane of the rectangles registered the current (see fig. 1).

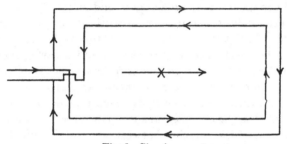

Fig. 1. Sizoo's experiment.

The total current must of necessity be $I_1 - I_2$, where $I_1$ is the current flowing in one rectangle and $I_2$ that in the other. At temperatures above the transition point the needle was deflected through an angle which could be computed from the potential and the relative cross-sections of the wires. When the temperature was lowered below the transition point, the deflection of the needle remained unchanged, showing that the current distribution is not altered on passing into the supra-conducting state. Moreover, if the current through the leads is now shut off, the needle nevertheless retains its position. Yet the current must now be flowing in the same direction in both circuits, which are connected in series and carrying a single current $I$ in two turns. Thus, as the deflection of the needle is not changed,

$$I = \frac{I_1 - I_2}{2}.$$

If the current is not turned on through the leads until the temperature is lowered below the transition point, the needle is not deflected at all. Thus in this case $I_1 = I_2$. When the temperature is raised above the transition point, the needle is deflected through the same angle as in the first experiment. Thus the magnetic energy of a supra-conducting circuit is constant and equal to the energy subsisting when the supra-conducting state was first attained.

## IV. II. 3.  *The Transition Curve*

Hitherto the magnetic field was considered only as an inducer of permanent currents. We must now consider the influence of a permanent magnetic field on a supra-conductor.

A magnetic field of sufficient strength destroys supra-conductivity. For a given supra-conductor and for every temperature below the normal transition point a certain field strength $H_k$ exists, known as the threshold field strength, which is just sufficient to cause resistance to appear in a supra-conductor. The threshold value $H_k$ is roughly proportional to $(T_t - T)$, where $T_t$ is the normal transition point for $H = 0$. Thus the curve showing the relation between $H_k$ and $T$, which is known as the transition curve, is approximately a straight line. The slope of the transition curve is very similar for most pure metals. In point of fact these curves are mostly slightly concave towards the origin.

The threshold value of the magnetic field generally shows a certain degree of hysteresis, i.e. is higher in increasing than in decreasing fields. In polycrystalline material this is usually confined to a slight indeterminacy in the value of $H_k$, the resistance appearing and vanishing not momentarily but in a narrow interval of field strength. But in the case of single crystals the effect is much more pronounced and depends on whether the field is parallel or perpendicular to the current. This is illustrated for the case of tin in figs. 2 and 3. In a longitudinal magnetic field the threshold value is sharp and reproducible in increasing fields; but when the field strength is reduced the resistance vanishes at more or less arbitrary field strengths, which are however always lower than those in rising fields. In a transversal field the resistance in rising fields begins to increase from zero at fields much lower than $H_k$, but it attains its normal value at $H_k$. In falling transversal

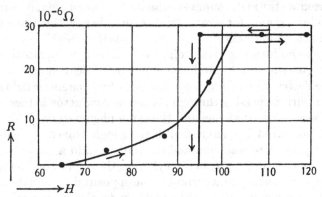

Fig. 2. Magnetic threshold curve of tin. $H \perp I$.

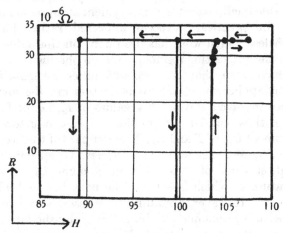

Fig. 3. Magnetic threshold curve of tin. $H \parallel I$.

fields the resistance vanishes sharply at a reproducible value of $H_k$. Similar phenomena occur in other pure supra-conductors.

Naturally enough the discovery of supra-conductivity raised the question of applying supra-conducting wires in electrical engineering. The advantages to be gained by employing wiring in which no Joule heat is produced are

obviously enormous. Unfortunately, apart from the difficulties connected with running electrical machinery at liquid
helium temperatures, it soon proved impossible to send large
currents through supra-conductors. For a given supra-
conductor and for every temperature below the transition
point a definite current $I_k$—the "critical current"—destroys
supra-conductivity. According to Silsbee $I_k$ is identical with
the current which produces a magnetic field equal to $H_k$.
The destruction of supra-conductivity by a magnetic field and
by a current passing through the supra-conductor is therefore
to be considered as one and the same phenomenon. This was
demonstrated by Tuyn and Kamerlingh Onnes as follows.
A copper wire was mounted coaxially within a tin cylinder.
When, on attaining $I_k$ in the cylinder, the supra-conductivity
of tin was destroyed, a current in the opposite direction through
the copper wire caused the resistance to disappear again.

The currents that may be passed through supra-conductors
are as a rule small, so that the employment of supra-conducting
metals for practical purposes appeared rather hopeless.
Nevertheless, much work has been done on this subject and
some rather interesting results have been obtained.

We have seen that the two difficulties inherent in the
technical application of supra-conductors are the low temperatures involved and the low values of $I_k$, i.e. of $H_k$. The
problem thus consists in finding supra-conductors with
maximum values of $T_l$ and $H_k$. The first part of this work was
attacked by Meissner with some success. Of all pure metals
the highest value of $T_l$ is found in niobium, the transition
temperature of which is 8·4° K. The next best is lead with
$T_l = 7\cdot26°$ K. However, according to Meissner's experiments,
a number of compounds are characterised by the fact that $T_l$
is considerably greater. Thus ZrN becomes supra-conducting
at 9·45°, TaC between 9·3° and 9·5° and NbC between 10·1°
and 10·5° K. This brings supra-conductivity into a range
of temperature that may be attained with solid hydrogen,
though it is still at present too low for practical application.
Of deeper significance for the subsequent development of
research are the results of de Haas and Voogt on the transition
curves of alloys. As in the case of most alloys the transition to

supra-conductivity does not occur at a definite field strength but in a considerable interval, the authors give as the critical field strength $H_k$ the value of $H_{\frac{1}{2}}$, i.e. the magnetic field which raises the resistance to one-half the normal value. This point has evidently no physical significance and was not very wisely chosen.

Whereas in pure supra-conductors a magnetic field of 100 gauss will destroy supra-conductivity at the lowest temperature attainable, a number of alloys will bear much higher field strengths. Thus for an alloy of the composition $Bi_5Tl_3$, which normally becomes supra-conducting at 6·4° K., at 4·2° $H_{\frac{1}{2}} = 4080$ gauss. At the same temperature a Pb-Hg-alloy, containing 15·3 per cent. of mercury, the normal transition point of which is 6·75°, will stand field strengths up to 6800 gauss. The most striking example is a Pb-Bi-alloy with 35 per cent. of bismuth. Its normal transition point is 8·7°. At 4·22° $H_{\frac{1}{2}} = 18,450$ gauss and at 1·91° $H_{\frac{1}{2}} = 26,700$ gauss. If we apply the approximate formula $H_k = a\,(T_l - T)$, the factor $a$, which say for tin equals 20, in the case of this alloy is about 4000.

It would thus appear that very considerable currents may be passed through alloys at temperatures slightly below their normal transition points. However, here difficulties of another nature have appeared, to which we shall refer in § 5.


### IV. II. 4. *The Supra-conducting State*

Hitherto we have considered a supra-conductor below its transition point as differing from the same conductor at higher temperatures merely in its infinitely small electrical resistance. The effect of the magnetic field in lowering the transition point was observed by measuring the resistance. Absence of resistance was the one and only criterion of what has often been termed the "supra-conducting state".

What are we to understand by this term? Is it merely a learned way of saying what we have said already, that under suitable conditions conductivity is infinite; or are we to suppose that in this state the supra-conductor has, apart from its lack of resistance, properties distinct from those it possesses

⟨ 275 ⟩                              18-2

in the ordinary condition? It is equivalent to asking whether any other properties undergo sudden changes on passing the transition point. This is one of the fundamental questions connected with supra-conductivity.

If the supra-conducting state were a "phase" in the usual thermodynamical sense, we should expect a latent heat at the transition point and a change in crystal structure. All attempts to discover the latter have failed hitherto, so that to decide the phase hypothesis one way or another it appeared necessary merely to determine whether a latent heat occurs or not. This was a typical *experimentum crucis*, as indicated in § 1, and it gave a typical answer. After Mendelssohn and Simon had been unable to detect any thermal effect in lead at the transition point, Keesom, van den Ende and Kok discovered in tin and thallium *not* a latent heat but a sudden drop in the specific heat curve (fig. 4). Lead with its low $\Theta$ and its high $T_t$ was an unfavourable example, as the normal specific heat at the transition point is too great to detect a small irregularity. Finally, when experiments were carried out on thallium in a constant magnetic field, a latent heat was indeed observed, but occurring not sharply at the transition point but spread over a narrow interval of temperature (fig. 5). The crucial experiment had merely added a complication.

Almost simultaneously a further property of the supra-conducting state was discovered. The thermal conductivity of a metal in this state differs from that of the same metal at the same temperature, in which supra-conductivity has been destroyed by a magnetic field. The thermal resistance-temperature curve, which is shown for the case of tin in fig. 6, forks at the transition point. The upper curve is obtained for $H = 0$ and for $H < H_k$; the lower curve for various values of $H > H_k$, where $H_k$ of course depends on the temperature. Strangely enough it is the upper, supra-conducting curve that appears to be the continuation of the course at higher temperatures. However, accurate experiments, carried out shortly afterwards on indium, showed that even without a magnetic field the thermal conductivity does indeed suffer a slight discontinuity at the normal transition point.

It is thus clear that the supra-conducting state is character-

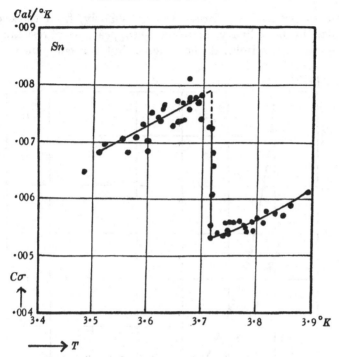

Fig. 4. Atomic heat of tin at the transition point.

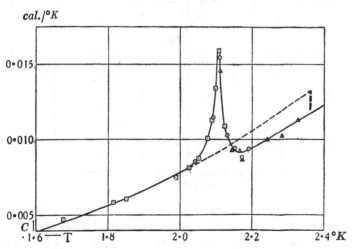

Fig. 5. Atomic heat of thallium in a magnetic field.

⟨ 277 ⟩

ised by properties differing, though slightly, from those of matter in the ordinary condition. Disappearance of resistance is no longer an isolated phenomenon but is accompanied by

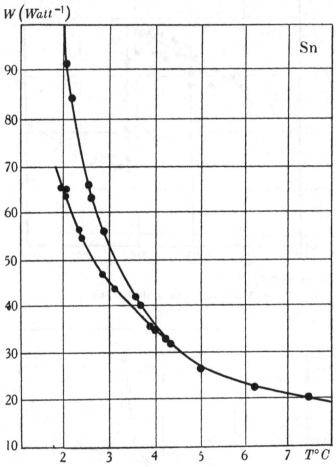

Fig. 6. Thermal resistance of tin. Upper curve: $H=0$; lower curve: $H>H_k$.

other changes. Though the supra-conducting state has evidently much in common with a thermodynamical phase, yet it is "not exactly a phase", and the phase hypothesis, of which we shall have something to say in the next paragraph, has still many difficulties to overcome. It is clear, moreover, that a magnetic field has a most fundamental influence on supra-conductors. Though its rôle in inducing permanent currents may be trivial, the threshold field $H_k$ has the elements of a variable determining an equilibrium curve, and we have just seen that a magnetic field can convert a drop in $C_p$ into a latent heat. In addition to this, in certain alloys the thermal resistance depends on the field even when $H < H_k$, i.e. within the supra-conducting state and not merely at the transition point. All this naturally leads us to ask what exactly are the magnetic properties of the supra-conducting state.

## IV. II. 5.  *Subpermeability*

According to the results of Sizoo's experiments, described in § 2 of this chapter, the permanent currents induced in supra-conducting circuits by magnetic fields maintain the magnetic flux constant through any closed surface bounded by a supra-conductor. Let us consider the magnetic lines of force in the case of two typical experiments. (i) A closed tin wire is cooled below its transition point and a magnetic field is then applied. A permanent current is thus induced, as a result of which the lines of force are prevented from penetrating a surface bounded by the wire (see fig. 7a). (ii) The circuit is cooled through its transition point in a constant magnetic field. On passing through $T_l$ no change of flux is observed and the lines of force are distributed evenly as in fig. 7b. No current flows through the circuit. Hereupon at $T < T_l$ the field is switched off, inducing a current that maintains the flux constant within the circuit, as shown in fig. 7c. Figs. 7 are purely schematical. In actual fact the distribution of the lines of force is considerably more intricate.

These phenomena are perfectly obvious and immediate consequences of the law of electromagnetic induction. They differ from those observed in the case of normal conductors merely in the fact that, owing to the infinite conductivity of the supra-conductor, the induced currents are not quenched. They can moreover be readily measured directly without considering the magnetic flux at all. We can, for instance, connect two points of the wire with the poles of a ballistic galvanometer and cut open the supra-conducting circuit. The galvanometer will then register a current impulse.

The phenomena become more complicated when we replace the closed wire by a supra-conducting ring of finite thickness

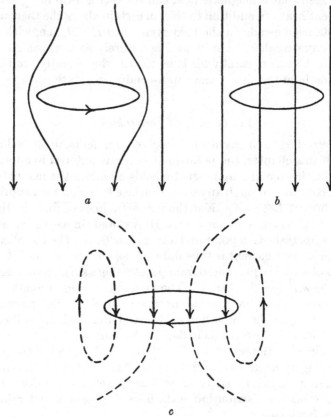

Fig. 7 *a, b, c.* Supra-conducting wire in a magnetic field.

or by a supra-conducting rod or sphere. Though the laws of inductance will be just the same in this case, we have now to consider the distribution of the magnetic flux in the supra-conducting material itself. We must in fact determine whether or not a magnetic field penetrates into the interior of a supra-conductor.

The traditional concept of persisting currents, together with the assumption that infinite conductivity is the only factor that influences the behaviour of a body in the supra-conducting state, led to the following conception of the magnetic properties of supra-conductors.

Let a supra-conductor below $T_l$ be placed in a magnetic field $H$. As long as $H < H_k$ the persisting currents will prevent a flux from arising in the conductor. Thus we shall have $B = 0$ (see fig. 8, I). When $H$ is increased to $H_k$, supra-conductivity

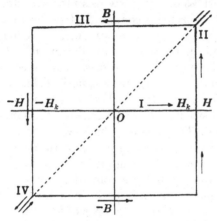

Fig. 8. Magnetic flux in a supra-conductor.

will be destroyed, the currents will be quenched and a normal flux $B = H_k$ will appear corresponding to $\mu = 1$, where $\mu$ is the permeability (II on fig. 8). A further increase in $H$ will lead to a corresponding increase in $B$ along the diagonal O—II, which will be traversed backwards when $H$ is decreased to $H_k$. When $H$ is further reduced below $H_k$ the induction flux $B$ will remain constant; $B = H_k$ until $H = -H_k$ (III, fig. 8). $B$ will then fall sharply to the diagonal at IV. Henceforward the figure will be symmetrical. Increase and decrease of $H$ will lead $B$ along the diagonal and around the rectangle as indicated by the arrows. This picture was first drawn explicitly by Schubnikow and Rjabinin in 1934, but it is implicitly contained in most concepts on supra-conductivity held up to that time. It entails two important corollaries: (1) The magnetic induction flux in a supra-conductor is not a unique function of the field strength but depends on the past history of the field and the specimen. (2) At field strengths $-H_k < H < H_k$ the induction flux is always $\pm H_k$ or 0 as long as the temperature remains constant throughout the experiment. No intermediate values are possible and an increase or decrease of $H$ within these limits will have no effect on $B$. Magnetic fields do not penetrate into supra-conductors.

The correctness of this view was rendered doubtful when Meissner and Ochsenfeld communicated the first result of a series of experiments which was being carried out simultaneously and independently at Berlin, Kharkov, Leiden, Oxford and Toronto. A lead rod was slowly cooled below its transition point in a constant magnetic field. The magnetic flux was measured with the help of a coil beside the rod. When the transition point was reached a sharp increase of flux was registered. This result, which was merely qualitative, is evidently incompatible with the picture we have drawn, and could be construed roughly as follows. On cooling a supra-conductor through its transition point in a magnetic field, the magnetic flux in the specimen does not remain constant and equal to $H_k$, but the lines of force are driven out of the supra-conductor, increasing the flux in its neighbourhood. Thus within the supra-conductor $B = 0$. This is illustrated in fig. 9. As fig. 9 is drawn for varying temperature and fig. 8 for varying field strength, the two are not strictly comparable. Nevertheless, the two pictures obviously disagree. According to one view $B = H_k$, according to the other, under the same conditions, $B = 0$.

Immediately after the publication of Meissner and Ochsenfeld's note, a large number of short communications appeared on results obtained with widely differing experimental procedures in various laboratories, but showing excellent agreement except in immaterial details. As most of these communications appeared in the form of letters to *Nature* it was difficult to evaluate the accuracy and full significance of each paper. At the time of writing most of these notes have been followed by detailed communications and we can now enumerate the various methods employed to determine the magnetic properties of a supra-conductor.

(1) At a constant temperature below $T_i$ the magnetic field of a solenoid, the lines of force of which are parallel to the axis of

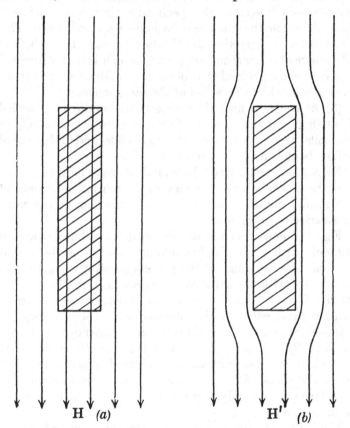

Fig. 9. Magnetic lines of force: (a) ordinary conductor; (b) supra-conductor.

a supra-conducting rod, is increased and decreased in small jerks. The change of induction is measured ballistically with the help of a coil wound round the specimen and connected to a galvanometer, which is suitably amplified. Similar experiments have been carried out on spheres and on supra-conductors of various shapes and dimensions.

(2) At constant temperature the rod or sphere is alternately introduced into and removed from a magnetic field of varying intensity. The induction is again measured ballistically, in this case the integral effect being obtained. Here the test coil is not wound directly on the specimen but a narrow gap is left, through which the body can be moved up and down. This necessitates a correction for the flux through the gap. The advantage of this method as against the first is that changes of flux can be determined that occur in periods of time as great as or greater than the period of the galvanometer.

(3) At constant field strength a supra-conductor is cooled through $T_t$ and the magnetic flux measured with the help of test coils placed at various positions outside, and in the case of hollow bodies, inside the specimen.

Numerous variations of these methods have been employed in different laboratories. In view of the mutual agreement of the results obtained we shall pick out a few of the most illustrative experiments.

Fig. 10 shows the results obtained by Schubnikow and Rjabinin with a rod of polycrystalline lead, using the second method. The beginning of the process (abcd) is completely in accordance with the old view as expressed in fig. 8. At first, for $H < H_k$, $B = 0$. At $H = H_k$, $B$ suddenly leaps to the normal value at $c$ and, with a further increase of $H$, proceeds along the diagonal ($d$). However, on the way back neither does $B$ remain equal to $H_k$ as in fig. 8, nor does it fall abruptly to zero, but sinks sharply to a definite value $e$ and, as $H$ is further reduced, gradually falls off. At $H = 0$ a residuary magnetisation remains, ($f$), which vanishes only at $H = -H_k$, ($g$), after which the curve becomes symmetrical.

This is the first time that a full magnetisation curve has been drawn for a supra-conductor, yet here again instead of deciding between two views the *experimentum crucis* has tendered a third. We may conclude from these experiments that, when the magnetic field around a polycrystalline supra-conductor is reduced below the threshold value, a certain part of the magnetic flux is immediately ejected, the remaining flux gradually following as the field strength is further reduced. $B = 0$ is not ultimately attained until $H = -H_k$. The residuary flux at *e* and *f* is not completely reproducible but depends on the conditions of the experiment and on the purity of the specimen. Moreover, further experiments carried out by Mendelssohn and his collaborators at Oxford and by Schubnikow at Kharkov proved that the residuary flux at *e* and *f* is different for different supra-conductors and depends on whether

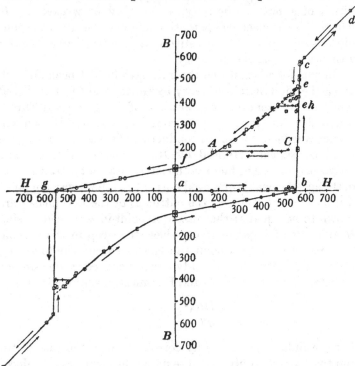

Fig. 10. Magnetic flux in polycrystalline lead.

the specimen is a single crystal or not. In the case of mercury no residual flux is found at $H = 0$; in a single crystal of lead the residuary flux is considerably smaller than that shown in fig. 10 at *e*. Moreover, this value decreases with time and could be shown to fall off continuously for half an hour at constant *H* and *T*. This brings Schubnikow to the conclusion that, at any rate in the case of single crystals of pure supraconducting metals, the only stable condition at $H < H_k$ is that in which $B = 0$.

Though the restriction to single crystals and pure metals is a drawback, the fact that the magnetic flux is equal to zero is a new and fundamental law of great significance. It enables us to characterise the supra-conducting state, in the restricted sense just mentioned, by the equation $B = 0$ or, which comes to the same thing, $\mu = 0$. This, supplementing the old condition of supra-conductivity, which may be written $E = 0$, where $E$ is the electrical potential difference between the ends of the supra-conductor, constitutes a practical foundation for a theory of supra-conductivity.

On this foundation two theories have in fact been erected, one a thermodynamical theory by Gorter and Casimir, based to a certain extent on considerations of Bridgman and Ehrenfest, and one an electromagnetical theory by F. and H. London. Gorter considers the supra-conducting state as a phase in the usual sense, characterised by $B = 0$, whereas the ordinary phase has $B = 1$. The fact that no latent heat but merely a drop in the specific heat is observed normally at the transition point can be accounted for, as the entropies of the two phases are shown to be equal at the transition point in zero field. Using $H$ and $T$ as independent variables, the drop in the specific heat at the transition point can be calculated. The following equation, which had already been deduced from similar considerations by Rutgers and Ehrenfest, is obtained:

$$\left(\frac{dH_k}{dT}\right)^2 = \frac{4\pi\Delta C}{Tv}.$$

This formula is satisfactorily corroborated in the case of tin and thallium, the only two pure metals hitherto tested. Moreover, the theory shows that at finite field strengths a latent heat is to be expected, as was actually found by Keesom and Kok.

F. and H. London make use of the condition $B = 0$ to deduce the electromagnetic equations of a supra-conductor. These have the form $\quad E = \Lambda\,(\dot{I} + c^2 \operatorname{grad}\rho)$

and $\qquad\qquad\qquad H = -\Lambda c \operatorname{curl} I,$

where $\rho$ is the current density and $\Lambda = m/ne^2$. These equations replace Ohm's law for a supra-conductor and, as against previous attempts at a theory, allow of electrostatic fields existing in supra-conductors. It is shown that permanent currents cannot in fact exist without the presence of a field, but that a magnetic field, which may in turn be produced by the current, is necessary in order that the latter may exist.

It appears quite clear that, if the condition $B = 0$ were really a characteristic property of the supra-conducting state, we should soon have a satisfactory picture of the phenomenon of supra-conductivity. Nevertheless, experiment is at present working in another direction, and the exceptions to this rule are now more the centre of interest than the rule itself. Keeley, Mendelssohn and Moore determined the amount of magnetic flux remaining in rods of various supra-conducting materials, when the magnetic field strength, which was originally greater than $H_k$, was gradually reduced to zero. For mercury 0 per cent. was found, for pure lead 15 per cent. in good agreement with Schubnikow, but 1 per cent. of bismuth added to lead produced a residuary flux at $H = 0$ of from 40 to 80 per cent., while lead with 10 per cent. bismuth gave 100 per cent. flux frozen in. Similarly, in alloys of tin and cadmium and of tin and bismuth the entire flux remained in the specimen down to $H = 0$.

These results are borne out by further experiments of Schubnikow and Rjabinin on the alloy $PbTl_2$. This is one of the alloys which can stand high field strengths without losing its supra-conductivity. At $2 \cdot 1°$ the threshold value is 1700 gauss, which we may call $H_{k_2}$. However, when $H$ is increased from zero, the magnetic field begins to penetrate at a very much smaller field strength $H_{k_1}$, which is equal to 100 gauss, a value compatible with the threshold fields of pure metals. When the field reaches this value, $B$ rises sharply, as shown in fig. 11, and then gradually approaches the diagonal, meeting it near $H_{k_2}$. When $H$ is thereupon decreased a slight hysteresis is observed, and at $H = 0$ a small residuary magnetisation remains, which does not depend on time. Similar phenomena were found in other alloys.

We are thus confronted by the fact that alloys, in contra-distinction to pure metals, appear to have two threshold values, one at which resistance appears, and another, much smaller, at which the field begins to penetrate into the conductor. Though this satisfactorily accounts for the influence of small magnetic fields on the thermal resistance of alloys, it makes the phenomenon of supra-conductivity even more complicated than before. Moreover, the hope of finding supra-conducting alloys through which large currents might be passed is moved still farther from fulfilment. Schubnikow and Rjabinin tested Silsbee's hypothesis on $PbTl_2$ and found that

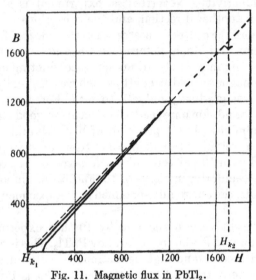

Fig. 11. Magnetic flux in $PbTl_2$.

the critical current, which destroys supra-conductivity, is not identical with the current which produces a field strength equal to the large value $H_{k_2}$, but is even slightly less than that which will produce the much smaller field strength $H_{k_1}$.

A subsequent full theory of supra-conductivity will thus have to account for two phenomena, the disappearance of resistance and the behaviour of the magnetic flux. Hitherto it is not yet apparent which of the two is the essential and primary effect, supra-conductivity or "subpermeability".

## NOTE ADDED IN PROOF

Since this book went to press new facts have been discovered which may do much to clear up the controversies still existing as regards the magnetic properties of supraconductors. It was shown by de Haas and Guineau that at field strengths above $2/3H_k$ a tin sphere becomes *uniformly* magnetised, its permeability gradually increasing with $H$ and attaining unity when $H = H_k$. We are thus confronted with what is apparently an intermediate state between supraconductivity and ordinary conductivity. In experiments of Schubnikow and Nakhutin the tin spheres remained supraconducting in the direction of the field up to fields equal to $H_k$, whereas perpendicular to the field supra-conductivity is destroyed near $2/3H_k$. This confirms a theory recently proposed by Landau that the intermediate state consists of a large number of layers alternately normally and supraconducting. The intermediate state and the anisotropic conductivity should be of fundamental importance for all future conceptions of supra-conductivity.

In other fields of research treated in this book, important developments have also occurred which could not be introduced into the text in proof. In order to make some amends for this we have added a list of references, which gives the literature that has appeared between the end of 1935 and May 1937.

# BIBLIOGRAPHY

\* Monograph or textbook.
† Contains description of apparatus.

## PART I

### CHAPTER I

\*G. CLAUDE. *Air liquide, oxygène, azote*. Paris, 1925.

\*J. R. VAN DER WALLS, jr. Die Zustandsgleichung. *Handbuch der Physik*, **10**, 1927.

L. CAILLETET. (Liquefaction of oxygen and air.) *C.R.* Dec. 24th, 1877.

H. KAMERLINGH ONNES. Methods and Apparatus used in the Cryogenic Laboratory, XIX. *Leiden Comm.* 158, 1926.

†K. OLSZEWSKI. Ueber die Dichte des flüssigen Methans, sowie des verflüssigten Sauerstoffs und Stickstoffs. *Ann. d. Phys. und Chem.* **31**, 58, 1887.

—— Ueber das Absorptionsspektrum des flüssigen Sauerstoffs und der verflüssigten Luft. *Wiener Berichte*, **2**, XCV, 1, 1887.

R. PICTET. (Liquefaction of oxygen.) *C.R.* Dec. 24th, 1877.

### CHAPTER II

\*H. HAUSEN. Die physikalischen Grundlagen der Gasverflüssigung und Rektifikation. *Geschichte der Gesellschaft für Lindes Eismaschinen*, p. 135, 1929.

\*H. LENZ. Gasverflüssigung und ihre thermodynamischen Grundlagen. *Handbuch der Experimentalphysik*, **9** (1), 47, 1929.

W. MEISSNER. Erzeugung tiefer Temperaturen und Gasverflüssigung. *Handbuch der Physik*, **11**, 1926.

### CHAPTER III

*Het natuurkundig laboratorium der Rijksuniversiteit te Leiden in de jaren* 1904–1922. Eduard Ijdo, Leiden, 1922.

J. DEWAR. (Liquefaction of hydrogen.) *Proc. Roy. Soc.* **63**, 256, 1898; *Journ. Chem. Soc. London*, **73**, 529, 1898.

†F. HENNING and A. STOCK. (Metal cryostat for temperatures from −150° to −190° C.) *Zs. f. Phys.* **4**, 226, 1921.

†L. HOLBORN and J. OTTO. (Metal cryostat for temperatures from 0° to −190°.) *Zs. f. Phys.* **30**, 320, 1924.

†H. KAMERLINGH ONNES. (Liquefaction of hydrogen.) *Leiden Comm.* 94*f*, 1906.

†—— (Liquefaction of helium.) *Leiden Comm.* 108, 1908.

BIBLIOGRAPHY

†H. Kamerlingh Onnes. (Cryostat for liquefied gases.) *Leiden Comm.* 94c, d, f.

†—— (Hydrogen vapour cryostat.) *Leiden Comm.* 151a.

†H. Kamerlingh Onnes and C. A. Crommelin. (Hydrogen vapour cryostat.) *Leiden Comm.* 154c.

†P. Kapitza. Hydrogen Liquefaction Plant at the Royal Society Mond Laboratory. *Nature*, Feb. 13th, 1932.

†—— The Liquefaction of Helium by an Adiabatic Method. *Proc. Roy. Soc.* (A), 1934.

†W. Meissner. (Cryostat for pressures above 1 atm.) *Ann. d. Phys.* 47, 1024, 1915.

†—— Heliumverflüssigungsanlage der Physikalisch-technischen Reichsanstalt. *Phys. Zs.* 26, 689, 1925.

†K. Mendelssohn. Ein Apparat nach dem Desorptionsverfahren für Messungen bis zu 2° abs. *Zs. f. Phys.* 73, 482, 1931.

†M. Ruhemann. Ein kleiner Apparat zur Erzeugung sehr tiefer Temperaturen. *Zs. f. Phys.* 65, 67, 1930.

†F. Simon. Ein neues einfaches Verfahren zur Erzeugung sehr tiefer Temperaturen. *Zs. f. d. gesamte Kälteindustrie*, 34, 1, 1927.

†—— Heliumverflüssigung mit Arbeitsleistung. *Zs. f. d. gesamte Kälteindustrie*, 39, 89, 1932.

†F. Simon and J. E. Ahlberg. Ein Demonstrationsapparat für Heliumverflüssigung. *Zs. f. Phys.* 81, 816, 1933.


CHAPTER IV

*F. Henning. *Die Grundlagen, Methoden und Ergebnisse der Temperaturmessung.* Braunschweig, 1915.

*H. Kamerlingh Onnes and W. H. Keesom. Die Zustandsgleichung. *Leiden Comm.* Suppl. 23.

J. G. Aston, E. Willihnganz and G. H. Messerly. A Thermodynamic Temperature Scale in terms of the Copper Constantan Thermocouple from 12° to 273° K. *Journ. Amer. Chem. Soc.* 57, 1643, 1935.

G. Borelius, W. H. Keesom, C. H. Johanson and J. O. Linde. Measurements of Thermoelectric Forces down to Temperatures obtainable with Liquid Helium. *Leiden Comm.* 217e, 1932.

K. Clusius. (Lead resistance thermometer.) *Zs. f. phys. Chem.* (B) 33, 41, 1929.

W. F. Giauque, R. M. Buffington and W. A. Schulze. Copper-constantan Thermocouple. *Journ. Amer. Chem. Soc.* 49, 2353, 1927.

W. F. Giauque and R. Wiebe. (Gold resistance thermometer.) *Journ. Amer. Chem. Soc.* 50, 101, 1928.

L. Holborn and F. Henning. Vergleichung von Platinthermometern mit dem Stickstoff-, Wasserstoff-, und Heliumthermometer. *Ann. d. Phys.* (4), 35, 761, 1911.

H. Kamerlingh Onnes and G. Holst. (Resistance of constantan and manganin.) *Leiden Comm.* 142a.

BIBLIOGRAPHY

W. H. KEESOM. Sur la base expérimentale, sur laquelle repose l'échelle internationale de température, en ce qui concerne des basses températures. *Leiden Comm.* Suppl. 67 *b*, 1929.

W. H. KEESOM and J. N. VAN DEN ENDE. Resistance Thermometry at the Temperatures of Liquid Helium. *Leiden Comm.* 203 *c*, 1929.

CHAPTER V

*J. R. VAN DER WAALS and P. KOHNSTAMM. *Lehrbuch der Thermostatik*, vol. II. Barth, Leipzig, 1927.

*P. KOHNSTAMM. Thermodynamik der Gemische. *Handbuch der Physik*, **10**, 223, 1926.

B. F. DODGE and A. K. DUNBAR. Coexisting Liquid and Vapour Phases of Oxygen-Nitrogen Solutions. *Journ. Amer. Chem. Soc.* **49**, 491, 1927.

V. FISCHER. Ein Zustandsdiagram für ternäre Gemische. *Ann. der Phys.* (5), **17**, 421, 1933.

H. HAUSEN. Die physikalischen Grundlagen der Gasverflüssigung und Rektifikation. *Geschichte der Gesellschaft für Linde's Eismaschinen*, 1929.

—— Verlustfreie Zerlegung von Gasgemischen durch umkehrbare Rektifikation. *Zs. f. techn. Phys.* **13**, 271, 1932.

—— Einfluss des Argons auf die Rektifikation der Luft. *Forschung auf dem Gebiet des Ingenieurwesens*, **5**, 290, 1934.

—— Rektifikation von Dreistoffgemischen insbesondere von Sauerstoff-Stickstoff-Argon-Gemischen. *Forschung auf dem Gebiet des Ingenieurwesens*, **6**, 9, 1935.

W. H. KEESOM. Étude sur la représentation graphique du processus de la rectification. *Sixième congrès international du froid*, Buenos Aires, pp. 136, 161, 1932; *Leiden Comm.* Suppl. 69, 71 *c*, 72 *a*.

H. LENZ. Gasverflüssigung und ihre thermodynamischen Grundlagen. *Handbuch der Experimentalphysik*, **9** (1), 1929.

†W. MEISSNER and K. STEINER. Verbesserter Apparat zur Trennung von Neon-Helium-Gemischen. *Zs. f. d. gesamte Kälteindustrie*, **39**, 49, 75, 1932.

W. MERKEL. (Enthalpy-composition diagram for oxygen-nitrogen mixtures.) *Archiv für Wärmewirtschaft*, **10**, 13, 1929.

K. PETERS and K. WEYL. Adsorptionsversuche an schweren Edelgasen. *Zs. f. phys. Chem.* (A), **148**, 1, 1930.

A. WEISSBERG. (Calculation of apparatus for the separation of air.) *Journal of Technical Physics* (Russian), **4**, 1204, 1934.

I. YUSHKJEWICH and N. ISHKIN. *The Technology of Bound Nitrogen. I. Production of Nitrogen and Oxygen from Air.* (Russian.) Moscow, 1934.

CHAPTER VI

*G. TAMMANN. *Aggregatzustände.* Leipzig, 1922.

*P. W. BRIDGMAN. *The Physics of High Pressures.* Bell and Sons, London, 1931.

—— Mercury Liquid and Solid under Pressure. *Proc. Amer. Acad.* **47**, 347, 1912.

P. W. BRIDGMAN. Change of Phase under Pressure. *Phys. Rev.* **3**, 126, 153, 1914; **6**, 1, 94, 1915.

—— The Melting Parameters of Nitrogen and Argon under Pressure and the Nature of the Melting-Curve. *Phys. Rev.* **46**, 930, 1934.

W. VAN GULIK and W. H. KEESOM. The Melting Curve of Hydrogen to 245 kg./cm.$^2$ *Leiden Comm.* 192 b, 1928.

†H. KAMERLINGH ONNES and W. VAN GULIK. The Melting Curve of Hydrogen to 55 kg./cm.$^2$ *Leiden Comm.* 184 a, 1926.

W. H. KEESOM. Solid Helium. *Leiden Comm.* 184 b, 1926.

W. H. KEESOM and J. H. C. LISMAN. The Melting Curve of Hydrogen to 450 kg./cm.$^2$ *Leiden Comm.* 213 e, 1931.

—— —— La courbe de fusion de l'hydrogène jusqu'à 610 kg./cm.$^2$ *Leiden Comm.* 221 a, 1932.

—— —— The Melting Curve of Neon to 200 kg./cm.$^2$ *Leiden Comm.* 224 b, 1933.

—— —— The Melting Curve of Nitrogen to 110 kg./cm.$^2$ *Leiden Comm.* 232 b, 1934.

J. H. C. LISMAN. Smeltlijnen van gecondenseerde Gassen. Leiden Thesis, 1934.

M. RUHEMANN and A. LICHTER. Zustandsdiagramme niedrig schmelzender Gemische, I. *Sow. Phys.* **6**, 139, 1934.

M. RUHEMANN, A. LICHTER and P. KOMAROW. Zustandsdiagramme niedrig schmelzender Gemische, II. *Sow. Phys.* **8**, 326, 1935.

F. SIMON and G. GLATZEL. Bemerkungen zur Schmelzdruckkurve. *Zs. f. anorg. Chem.* **178**, 309, 1929.

†F. SIMON, M. RUHEMANN and W. A. M. EDWARDS. Untersuchung über die Schmelzkurve des Heliums, I, II. *Zs. f. phys. Chem.* (B), **2**, 340; **6**, 62, 1929.

—— —— Die Schmelzkurven von Wasserstoff, Neon, Stickstoff und Argon. *Zs. f. phys. Chem.* (B), **6**, 331, 1930.

†T. T. H. VERSCHOYLE. The Ternary System—carbon monoxide, nitrogen, hydrogen. *Trans. Roy. Soc.* (A), **230**, 189, 1931.

# PART II

## CHAPTER I

*P. P. EWALD and C. HERMANN. Strukturbericht, 1913–1928. *Akademische Verlagsgesellschaft.* Leipzig, 1931.

*W. H. KEESOM. Rapport sur les recherches concernant la structure des substances dans les états solides et liquides aux basses températures, faites entre le 4ième et 5ième congrès du froid. *Leiden Comm.* Suppl. 64, 1928.

D. CALLIHAN and E. O. SALANT. Modified Scattering by Crystallised HCl and HBr. *Journ. Chem. Phys.* **2**, 317, 1934.

†K. CLUSIUS. Ueber die spezifische Wärme einiger kondensierter Gase zwischen 10° abs. und ihrem Tripelpunkt (N$_2$, O$_2$, CO, CH$_4$, HCl). *Zs. f. phys. Chem.* (B), **3**, 41, 1929.

K. CLUSIUS. Freie Rotation im Gitter des Monosilans. *Zs. f. phys. Chem.* (B), **23**, 213, 1934.

†K. CLUSIUS and A. PERLICK. Die Unstetigkeit im thermischen und kalorischen Verhalten des Methans bei 20·4° abs., als Phasenumwandlung zweiter Ordnung. *Zs. f. phys. Chem.* (B), **24**, 313, 1934.

R. M. CONE, G. H. DENNISON and J. D. KEMP. The Dielectric Constant of HCl from 85° to 165° K. *Journ. Amer. Chem. Soc.* **53**, 1278, 1931.

J. L. CRENSHAW and I. RITTER. Spezifische Wärmen einiger Ammoniumsalze. *Zs. f. phys. Chem.* (B), **16**, 143, 1932.

†H. EPSTEIN and W. STEINER. Messungen des Ramaneffekts bei tiefen Temperaturen ($C_6H_6$, HI). *Zs. f. phys. Chem.* (B), **26**, 131, 1934.

A. EUCKEN and E. KARWAT. Die Bestimmung des Wärmeinhalts einiger kondensierter Gase (HCl, HBr, HI, $NH_3$, $CH_4$, NO). *Zs. f. phys. Chem.* **112**, 467, 1927.

R. H. FOWLER. A Theory of the Rotation of Molecules in Solids and of the Dielectric Constant of Solids and Liquids. *Proc. Roy. Soc.* (A), **149**, 1, 1935.

—— The Anomalous Specific Heats of Crystals with Special Reference to the Contribution of Molecular Rotations. *Proc. Roy. Soc.* (A), **151**, 1, 1935.

W. F. GIAUQUE and R. WIEBE. The Entropy of HCl. Heat Capacity from 16° K. to the Boiling Point. *Journ. Amer. Chem. Soc.* **50**, 101, 1928.

S. B. HENDRICKS. Die Kristallstruktur von $N_2O_4$. *Zs. f. Phys.* **70**, 699, 1931.

S. B. HENDRICKS, E. POZNJAK and F. C. KRACEK. Molecular Rotation in the solid State. The Variation of the Crystal Structure of Ammonium Nitrate with Temperature. *Journ. Amer. Chem. Soc.* **54**, 2766, 1932.

C. HERMANN and M. RUHEMANN. Die Kristallstruktur von Quecksilber. *Zs. f. Krist.* **83**, 136, 1932.

R. HETTICH. Ueber den Tieftemperaturzustand der Ammoniumsalze (Piezoeffekt). *Zs. f. phys. Chem.* **168**, 353, 1934.

†G. HETTNER. Eine Doppelbande des festen Chlorwasserstoffs. *Zs. f. Phys.* **78**, 141, 1933.

†—— Die Kernschwingungsbande des festen und flüssigen Chlorwasserstoffs zwischen 20° und 160° abs. *Zs. f. Phys.* **89**, 234, 1934.

†G. HETTNER and F. SIMON. Ultrarotspektrum von Ammoniumsalzen im Umwandlungsgebiet. *Zs. f. phys. Chem.* (B), **1**, 293, 1928.

J. D. HITCHCOCK and C. P. SMYTH. Rotation of Molecules or Groups in Crystalline Solids. *Journ. Amer. Chem. Soc.* **55**, 1296, 1933.

W. H. KEESOM and I. W. L. KÖHLER. New Determination of the lattice Constant of $CO_2$. *Leiden Comm.* 230, 1934.

—— —— The Lattice Constant and Expansion Coefficient of Solid $CO_2$. *Leiden Comm.* 232, 1934.

W. H. KEESOM and H. H. MOOY. The Crystal Structure of Krypton. *Nature*, **125**, 889, 1930.

W. H. KEESOM, J. DE SMEDT and H. H. MOOY. The Crystal Structure of Parahydrogen at Liquid Helium Temperature. *Proc. Acad. Amst.* **33**, 814, 1930.

BIBLIOGRAPHY

W. H. KEESOM and K. W. TACONIS. An X-ray Goniometer for the Investigation of the Crystal Structure of Solidified Gases (Ethylene). *Physica*, 2, 463, 1935.

S. E. KEMERLING and C. P. SMYTH. Dipole Rotation in certain Crystalline Solids. *Journ. Amer. Chem. Soc.* 55, 462, 1933.

J. D. KEMP and G. H. DENNISON. Dielectric Constant of Solid $H_2S$. *Journ. Amer. Chem. Soc.* 55, 251, 1933.

J. A. A. KETELAAR. The Crystal Structure of the Low Temperature Modification of Ammonium Bromide (and Iodide). *Nature*, 134, 250, 1934.

F. C. KRACEK, E. POSNJAK and F. G. HENDRICKS. Gradual Transition in Sodium Nitrate. II. The Structure at Various Temperatures and its Bearing on Molecular Rotation. *Journ. Amer. Chem. Soc.* 53, 3339, 1931.

W. E. LASHKAREW and J. D. USYSKIN. Die Bestimmung der Lage der Wasserstoffionen im $NH_4Cl$-Gitter durch Elektronenbeugung. *Zs. f. Phys.* 85, 618, 1933.

A. C. MENZIES and H. R. MILLS. Raman Effect and Temperature. I. $NH_4Cl$, $NH_4Br$ and $NH_4I$. *Proc. Roy. Soc.* (A), 148, 407, 1935.

H. H. MOOY. On the Crystal Structure of Methane. I. *Leiden Comm.* 213, 1931.

—— On the Crystal Structure of Methane. II. *Leiden Comm.* 216, 1931.

—— Note relative à des recherches préliminaires aux rayons-X sur l'oxygène, l'acétylène et l'éthylène à l'état solide. *Leiden Comm.* 223, 1932.

G. NATTA. The Crystal Structure and Polymorphism of Hydrogen Halides. *Nature*, 127, 235, 1931.

G. NATTA and G. NASINI. The Crystal Structure of Xenon. *Nature*, 125, 457, 1930.

—— —— The Crystal Structure of Krypton. *Nature*, 125, 889, 1930.

H. H. NIELSEN. The Rotation of Molecules in Crystals. *Journ. Chem. Phys.* 3, 189, 1935.

L. PAULING. The Rotational Motion of Molecules in Crystals. *Phys. Rev.* 36, 430, 1930.

R. POHLMANN. Ultrarotspektren von Ammoniumsalzen im Gebiet ihrer anomalen spezifischen Wärme. *Zs. f. Phys.* 79, 394, 1932.

†B. RUHEMANN. Eine neue Röntgenkamera für tiefe Temperatur. *Sow. Phys.* 7, 572, 1935.

†B. RUHEMANN and F. SIMON. Die Kristallstrukturen von Krypton, Xenon, Bromwasserstoff und Jodwasserstoff in ihrer Abhängigkeit von der Temperatur. *Zs. f. phys. Chem.* (B), 15, 389, 1932.

†M. RUHEMANN. Röntgenographische Untersuchungen an festem Stickstoff und Sauerstoff. *Zs. f. Phys.* 76, 368, 1932.

P. E. SHEARER. The Infra-red Absorption Spectrum of Solid Hydrogen Chloride. *Phys. Rev.* 48, 299, 1935.

†F. SIMON and C. VON SIMSON. Die Kristallstruktur des Chlorwasserstoffs. *Zs. f. Phys.* 21, 168, 1924.

—— —— Die Kristallstruktur des Argons. *Zs. f. Phys.* 25, 160, 1924.

—— —— Umwandlungspunkte der Ammoniumsalze zwischen −30° und −40° C. *Naturw.* 14, 880, 1926.

BIBLIOGRAPHY

†F. Simon and E. Vohsen. Kristallstrukturbestimmung der Alkalimetalle und des Strontiums. *Zs. f. phys. Chem.* **133**, 165, 1928.

J. de Smedt and W. H. Keesom. The crystal structure of Argon. Researches on the crystal structure of nitrogen and oxygen at the temperature of liquid hydrogen. *Leiden Comm.* 178, 1925.

J. de Smedt, W. H. Keesom and H. H. Mooy. Analyse crystalline de l'azote solide. I. *Leiden Comm.* 202, 1929.

†—— —— —— On the Crystal Structure of Neon. *Leiden Comm.* 203, 1930.

C. P. Smyth and C. S. Hitchcock. Dipole Rotation in Crystalline Solids. *Journ. Amer. Chem. Soc.* **54**, 4631, 1932.

†—— —— Dipole Rotation and the Transitions in Crystalline Hydrogen Halides. *Journ. Amer. Chem. Soc.* **55**, 1830, 1933.

T. E. Sterne. The Symmetric Spherical Oscillator and the Rotational Motion of Homopolar Molecules in Crystals. *Proc. Roy. Soc.* (A), **130**, 551, 1931.

G. B. B. M. Sutherland. Experiments on the Raman Effect at very Low Temperatures. *Proc. Roy. Soc.* (A), **141**, 535, 1933.

†L. Vegard. Die Struktur derjenigen Form von festem $N_2$, die unterhalb von 35·5° K. stabil ist. *Zs. f. Phys.* **58**, 497, 1929.

—— Struktur und Leuchtfähigkeit von festem Kohlenoxyd. *Zs. f. Phys.* **61**, 185, 1930.

†—— Die Struktur von festem $N_2O_4$ bei der Temperatur der flüssigen Luft. *Zs. f. Phys.* **68**, 184, 1931.

—— Die Kristallstruktur von $N_2O_4$. *Zs. f. Phys.* **71**, 299, 1931.

†—— Die Struktur von festem $H_2S$ und $H_2Se$ bei der Temperatur der flüssigen Luft. *Zs. f. Krist.* **77**, 23, 1931.

†—— Die Struktur von β-$N_2$ usw. *Zs. f. Phys.* **79**, 471, 1932.

—— Die Struktur der β-Form des festen Kohlenoxyds. *Zs. f. Phys.* **88**, 235, 1934.

†L. Wilberg. Das Verhalten der Kernschwingungsbanden des Ammonium-radikals im Umwandlungsgebiet. *Zs. f. Phys.* **64**, 304, 1930.

CHAPTER II

*A. Eucken. Energie und Wärmeinhalt. *Handbuch der Experimentalphysik.*
*E. Schrödinger. Spezifische Wärme (Theoretischer Teil). *Handbuch der Physik*, **10**, 275, 1926.

J. B. Austin and R. H. H. Pierce, jr. Linear Thermal Expansion of a single Crystal of $NaNO_3$. *Journ. Amer. Chem. Soc.* **55**, 661, 1933.

H. A. Bethe. Statistical Theory of Superlattices. *Proc. Roy. Soc.* (A), **150**, 552, 1935.

J. M. Bijvoet and J. A. A. Ketelaar. Molecular Rotation in Solid $NaNO_3$. *Journ. Amer. Chem. Soc.* **54**, 1507, 1932.

M. Blackman. Contributions to the Theory of the Specific Heats of Crystals. I and II. *Proc. Roy. Soc.* (A), **148**, 365, 384, 1935.

—— Contributions to the Theory of the Specific Heat. III and IV. *Proc. Roy. Soc.* (A), **149**, 117, 126, 1935.

⟨ 297 ⟩

W. BRAGG and E. G. WILLIAMS. The Effect of Thermal Agitation on the Atomic Arrangement in Alloys. *Proc. Roy. Soc.* (A), **145**, 699, 1934.

K. CLUSIUS. Ueber Umwandlungen in festen Gasen ($SiH_4$, $PH_3$, $SH_2$). *Zs. f. Elektrochemie*, **39**, 598, 1933.

†K. CLUSIUS and P. HARTECK. Ueber die spezifischen Wärmen einiger fester Körper bei tiefen Temperaturen. (Gold, zinc, gallium, zinc oxide, copper oxide, zinc sulphide, ammonium carbaminate, silver chloride.) *Zs. f. phys. Chem.* (A), **134**, 243, 1928.

K. CLUSIUS, K. HILLER and J. V. VAUGHEN. Ueber die spezifischen Wärmen des Stickoxyduls, des Ammoniaks und Fluorwasserstoffs von 10° K. *Zs. f. phys. Chem.* (B), **8**, 427, 1930.

†K. CLUSIUS and A. PERLICK. Die Unstetigkeit im thermischen und kalorischen Verhalten des Methans bei 20·4° als Phasenumwandlung zweiter Ordnung. *Zs. f. phys. Chem.* (B), **24**, 313, 1934.

†J. L. CRENSHAW and I. RITTER. Spezifische Wärmen einiger Ammoniumsalze. *Zs. f. phys. Chem.* (B), **16**, 143, 1932.

†S. CRISTESCU and F. SIMON. Die spezifischen Wärmen von Beryllium, Germanium und Hafnium bei tiefen Temperaturen. *Zs. f. phys. Chem.* (B), **25**, 273, 1934.

U. DEHLINGER. Ueber Umwandlungen von festen Metallphasen. *Zs. f. Phys.* **83**, 832, 1933.

—— Stetiger Uebergang und kritischer Punkt zwischen zwei festen Phasen. *Zs. f. phys. Chem.* (B), **26**, 343, 1934.

—— Ueber die Existenz einer Umwandlung von genau zweiter Ordnung. *Zs. f. phys. Chem.* (B), **28**, 112, 1935.

P. EHRENFEST. Phasenumwandlungen im üblichen und erweiterten Sinn klassifiziert nach den entsprechenden Singularitäten des thermodynamischen Potentials. *Leiden Comm.* Suppl. 75, 1933.

†A. EUCKEN and H. WERTH. Die spezifische Wärmen einiger Metalle und Metallegierungen bei tiefen Temperaturen (Cu, Ni, constantan, Fe, Fe-Mn). *Zs. f. anorg. Chem.* **188**, 152, 1930.

†W. F. GIAUQUE and R. WIEBE. The Entropy of HCl, Heat Capacity from 16° K. to Boiling Point. *Journ. Amer. Chem. Soc.* **50**, 101, 1928.

—— —— The Heat Capacity of Hydrogen Bromide from 15° K. to its Boiling Point. *Journ. Amer. Chem. Soc.* **50**, 2193, 1928.

—— —— The Heat Capacity of HI from 15° K. to its Boiling Point. *Journ. Amer. Chem. Soc.* **51**, 1441, 1929.

W. S. GORSKI. Röntgenographische Untersuchung von Umwandlungen in der Legierung Cu-Au. *Zs. f. Phys.* **50**, 64, 1928.

†R. KAISCHEW. Thermische Untersuchungen an festem und flüssigem Helium. Breslau Thesis, 1932.

R. KAISCHEW and F. SIMON. Some Thermal Properties of Condensed Helium. *Nature*, **133**, 460, 1934.

H. KAMERLINGH ONNES and W. H. KEESOM. Specific heats of copper and lead. *Leiden Comm.* 143.

—— —— Quelques remarques en rapport avec l'anomalie de la chaleur spécifique de l'hélium liquide au point lambda. *Leiden Comm.* Suppl. 71, 1932.

W. H. KEESOM. On the Jump in the Expansion Coefficient of Liquid Helium in passing the λ-point. *Leiden Comm.* Suppl. 75, 1933.

W. H. KEESOM and C. W. CLARK. The Atomic Heat of Nickel from 1·1 to 19° K. *Physica,* 2, 513, 1935.

―― ―― The Heat Capacity of Potassium Chloride from 2·3 to 17° K. *Physica,* 2, 698, 1935.

W. H. KEESOM and K. CLUSIUS. Die Umwandlung flüssiges Helium I—flüssiges Helium II unter Druck. *Leiden Comm.* 216, 1931.

―― ―― Ueber die spezifische Wärme des flüssigen Heliums. *Leiden Comm.* 219, 1932.

W. H. KEESOM and J. N. VAN DEN ENDE. The Specific Heat of Solid Substances at the Temperatures of Liquid Helium. II (lead and bismuth). *Leiden Comm.* 203, 1930.

―― ―― The Specific Heat of Solid Substances at the Temperatures of Liquid Helium. III (lead and bismuth, correction). *Leiden Comm.* 213, 1931.

†―― ―― The Specific Heats of Solid Substances at the Temperatures of Liquid Helium. IV (tin and zinc). *Leiden Comm.* 219, 1932.

W. H. KEESOM and A. P. KEESOM. On the Anomaly in the Specific Heat of Liquid Helium. *Leiden Comm.* 221, 1932.

†―― ―― Isopyknals of Liquid Helium. *Leiden Comm.* 224, 1933.

―― ―― Thermodynamic Diagrams of Liquid Helium. *Leiden Comm.* Suppl. 76, 1933.

―― ―― The Entropy Diagram of Liquid Helium. *Leiden Comm.* Suppl. 76, 1933.

†―― ―― New Measurements on the Specific Heat of Liquid Helium. *Physica,* 2, 557, 1935.

W. H. KEESOM and J. A. KOK. On a Method of Correcting for incomplete Thermal Insulation in Measurements of Small Heat Capacities. *Leiden Comm.* 219, 1932.

―― ―― Measurements of the Specific Heats of Silver. *Leiden Comm.* 219, 1932.

―― ―― On the Change of the Specific Heat of Tin in becoming Supraconductive. *Leiden Comm.* 221, 1932.

―― ―― Measurements of the Specific Heat of Thallium at Liquid Helium Temperatures. *Leiden Comm.* 230, 1934.

―― ―― Measurements of the Latent Heat of Thallium Connected with the Transition, in a Constant Magnetic Field, from the Supra-con ductive to the non-Supra-conductive State. *Leiden Comm.* 230, 1934.

―― ―― Further Calorimetric Experiments on Thallium. *Leiden Comm.* 232, 1934.

K. K. KELLEY. Heat Capacity of Methyl Alcohol from 16 to 298° K. and corresponding Entropy and Free Energy. *Journ. Amer. Chem. Soc.* 51, 180, 1929.

F. KRACEK. Gradual Transition in Sodium Nitrate. I. Physico-chemical Criteria of the Transition. *Journ. Amer. Chem. Soc.* 53, 2609, 1931.

# BIBLIOGRAPHY

F. C. Kracek, E. Posnjak and S. B. Hendricks. Gradual Transition in Sodium Nitrate. II. The Structure at various Temperatures and its Bearing on Molecular Rotation. *Journ. Amer. Chem. Soc.* **53**, 333, 1931.

†N. Kürti. Ueber das thermische und magnetische Verhalten des Gadoliniumsulfats im Temperaturgebiet des flüssigen Heliums. *Zs. f. phys. Chem.* (B), **20**, 305, 1933.

L. Landau. Zur Theorie der Anomalie der spezifischen Wärme. *Sow. Phys.* **8**, 113, 1935.

†F. Lange. Untersuchung über die spezifische Wärme bei tiefen Temperaturen (W, Sn white and grey, I). *Zs. f. phys. Chem.* **110**, 343, 1924.

F. Lange and F. Simon. Spezifische Wärme und chemische Konstante des Kadmiums. *Zs. f. phys. Chem.* **134**, 374, 1928.

K. Mendelssohn and J. O. Closs. Kalorimetrische Untersuchungen im Temperaturgebiet des flüssigen Heliums (Ag, Cu). *Zs. f. phys. Chem.* (B), **19**, 291, 1932.

K. Mendelssohn, M. Ruhemann and F. Simon. Die spezifischen Wärmen des festen Wasserstoffs bei Helium-temperaturen. *Zs. f. phys. Chem.* (B), **15**, 121, 1931.

R. W. Millar. Specific Heats at Low Temperatures of MnO, $Mn_3O_4$, and $MnO_2$. *Journ. Amer. Chem. Soc.* **50**, 1875, 1928.

G. W. Pankow and P. Scherrer. Anomalie der spezifischen Wärme des Lithiums. *Helv. Phys. Acta*, **7**, 644, 1934.

B. Ruhemann. Temperaturabhängigkeit der Gitterkonstanten von Manganoxyd. *Sow. Phys.* **7**, 590, 1935.

†F. Simon. Untersuchungen über die spezifischen Wärmen bei tiefen Temperaturen (glasses, Hg, $NH_4Cl$, glycerine). *Ann. der Phys.* **68**, 241, 1922.

—— Thermisch erregte Quantensprünge in festen Körpern. *Berl. Ber.* **33**, 477, 1926.

—— Zum Prinzip von der Unerreichbarkeit des absoluten Nullpunktes. *Zs. f. Phys.* **41**, 806, 1927.

—— Interpretation of Infra-red Frequencies of the Diamond. *Nature*, **125**, 855, 1930.

—— Behaviour of Condensed Helium near Absolute Zero. *Nature*, **133**, 529, 1934.

—— Application of Low Temperature Calorimetry to Radioactive Measurements. *Nature*, **135**, 762, 1935.

†F. Simon and R. Bergmann. Thermisch erregte Quantensprünge in festen Körpern. IV. Messung der thermischen Ausdehnung im Gebiet der Anomalie (Li, Si, Fe, Ni, Cu, $NH_4Cl$, $NH_4Br$, $NH_4PH_3$). *Zs. f. phys. Chem.* (B), **8**, 255, 1930.

†F. Simon and F. Lange. Die thermischen Daten des kondensierten Wasserstoffs. *Zs. f. Phys.* **15**, 312, 1923.

F. Simon, Cl. v. Simson and M. Ruhemann. Die spezifischen Wärmen der Ammoniumhalogenide zwischen −70° und Zimmertemperatur. *Zs. f. phys. Chem.* **129**, 339, 1927.

BIBLIOGRAPHY

F. SIMON and R. C. SWAIN. Untersuchungen über die spezifischen Wärmen bei tiefen Temperaturen (Li, Fe, $CaCO_3$, $Al_2O_3$, sorbed gases). *Zs. f. phys. Chem.* (B), **28**, 189, 1935.

F. SIMON and W. ZEIDLER. Untersuchungen über die spezifischen Wärmen bei tiefen Temperaturen (Na, K, Mo, Pt). *Zs. f. phys. Chem.* **123**, 383, 1926.

A. SMITS. Innere Gleichgewichte in den festen Phasen. I und II. *Phys. Zs.* **31**, 376, 1930; **35**, 914, 1934.

C. SYKES. Methods for Investigating Thermal Changes occurring during Transformations in a Solid Solution (β-brass). *Proc. Roy. Soc.* (A) **148**, 422, 1935.

G. TAMMANN. Ueber Umwandlungen in homogenen Stoffen. *Zs. f. phys. Chem.* (A), **170**, 380, 1934.

R. W. TAYLOR. Magnetic Susceptibility of MnO as a Function of Temperature. *Phys. Rev.* **44**, 776, 1933.

O. N. TRAPEZNIKOWA and L. W. SCHUBNIKOW. Ueber die Anomalie der spezifischen Wärme von wasserfreiem Eisenchlorid. *Sow. Phys.* **7**, 66, 1935.

—— —— Ueber die Anomalie der spezifischen Wärme von wasserfreiem $CrCl_3$. *Sow. Phys.* **7**, 255, 1935.

CHAPTER III

*K. BENNEWITZ. Der Nernst'sche Wärmesatz. *Handbuch der Physik*, **9**, 141, 1926.

*R. H. FOWLER and T. E. STERNE. Statistical Mechanics with Particular Reference to the Vapour Pressures and Entropies of Crystals. *Review of Modern Physics*, **4**, 635, 1932.

*W. NERNST. *Grundlagen des neuen Wärmesatzes.* Halle, 1918.

*F. SIMON. Die Bestimmung der freien Energie. *Handbuch der Physik*, **10**, 350, 1926.

*—— Fünfundzwanzig Jahre Nernst'sches Theorem. *Erg. der exakt. Naturw.* **9**, 1930.

A. EUCKEN and F. FRIED. Ueber die Nullpunktsentropie kondensierter Gase. *Zs. f. Phys.* **29**, 36, 1924.

A. EUCKEN, E. KARWAT and F. FRIED. Die Konstante I der thermodynamischen Dampfdruckgleichung bei mehratomigen Molekülen. *Zs. f. Phys.* **29**, 1, 1924.

F. LANGE. Untersuchungen über die spezifische Wärme bei tiefen Temperaturen (white and grey tin). *Zs. f. phys. Chem.* **110**, 343, 1924.

F. SIMON. Vom Prinzip von der Unerreichbarkeit des absoluten Nullpunktes. *Zs. f. Phys.* **41**, 806, 1927.

H. ZEISE. Spektralphysik und Thermodynamik. Die Berechnungen von freien Energien, Entropien, spezifischen Wärmen und Gleichgewichten aus spektroskopischen Daten und die Gültigkeit des dritten Hauptsatzes. *Zs. f. Elektrochemie*, **39**, 758, 895, 1933; **40**, 662, 885, 1934.

For the work of W. F. GIAUQUE and collaborators, cf. Table XIX and Part III, chap. I.

# PART III

## CHAPTER I

\*A. Farkas. *Orthohydrogen, Parahydrogen and Heavy Hydrogen.* Cambridge University Press, 1935.

H. Beutler. Stösse zweiter Art bei Molekülen. *Zs. f. Phys.* **50**, 581, 1928.

K. F. Bonhoeffer and P. Harteck. Para- und Orthowasserstoff. *Zs. f. phys. Chem.* (B), **4**, 113, 1929.

K. Clusius and E. Bartholomé. Die Rotationswärmen der Moleküle HD und D₂. *Naturw.* **22**, 297, 1934; *Zs. f. phys. Chem.* (B), **30**, 237, 258, 1935.

—— —— Zur Rotationswärme des schweren Orthowasserstoffs. *Zs. f. phys. Chem.* (B), **29**, 162, 1935.

—— —— Die Eigenschaften des kondensierten schweren Wasserstoffs. *Phys. Zs.* **35**, 969, 1934.

K. Clusius and A. Hiller. Spezifische Wärmen des Parawasserstoffs im festen, flüssigen und gasförmigen Zustand. *Zs. f. phys. Chem.* (B), **4**, 158, 1929.

D. M. Dennison. A Note on the Specific Heat of the Hydrogen Molecule. *Proc. Roy. Soc.* (A), **115**, 483, 1927.

G. von Elbe and F. Simon. Kalorimetrische Bestimmung des Energie-unterschiedes der beiden Wasserstoffmodifikationen. *Zs. f. phys. Chem.* (B), **6**, 79, 1929.

A. Eucken. (Specific heat of hydrogen gas.) *Berl. Ber.* 144, 1912; *Ber. deutsche phys. Ges.* **18**, 4, 1916.

A. Eucken and A. Hiller. Nachweis einer Umwandlung des Ortho-wasserstoffs in Parawasserstoff durch Messung der spezifischen Wärme. *Naturw.* **17**, 182, 1929; *Zs. f. phys. Chem.* (B), **4**, 142, 1929.

A. Farkas and L. Farkas. Experiments on Heavy Hydrogen. I. *Proc. Roy. Soc.* (A), **144**, 467, 1934.

A. Farkas, L. Farkas and P. Harteck. Experiments on Heavy Hydrogen. II. *Proc. Roy. Soc.* (A), **144**, 481, 1934.

W. F. Giauque. The Calculation of Free Energy from spectroscopic Data. *Journ. Amer. Chem. Soc.* **52**, 4808, 1930.

—— The Entropy of Hydrogen and the Third Law of Thermodynamics. *Journ. Amer. Chem. Soc.* **52**, 4816, 1930.

W. F. Giauque and H. L. Johnston. Symmetrical and Antisymmetrical Hydrogen and the Third Law of Thermodynamics. *Journ. Amer. Chem. Soc.* **50**, 3221, 1928.

K. Mendelssohn, M. Ruhemann and F. Simon. Die spezifischen Wärmen des festen Wasserstoffs bei Heliumtemperaturen. *Zs. f. phys. Chem.* (B), **15**, 121, 1931.

W. Schottky. *Phys. Zs.* **22**, 1, 1921; **23**, 9, 448, 1922.

## CHAPTER II

*P. Weiss and G. Foex. *Le Magnétisme.* Armand Colin, Paris, 1926.

*E. C. Wiersma. *Eenige Onderzoekingen over Paramagnetisme.* Martinus Nijhoff, 's Gravenhage, 1932.

*Papers by J. Becquerel and Leiden Collaborators on paramagnetic rotation in crystals. *Leiden Comm.* 103, 191c, 193a, 199a, b, 204a, b, 211a, c, 218a, 231a, Suppl. 20, 68a, b, c; *Zs. f. Phys.* 58, 205, 1929; *Physica,* 1, 383, 1934.

H. Becquerel. (Theory of magnetic rotation.) *C.R.* 125, 679, 1897.

J. W. Ellis and H. O. Kneser. Kombinationsbeziehungen im Absorptionsspektrum des flüssigen Sauerstoffs. *Zs. f. Phys.* 86, 583, 1933.

G. Foex. (Cryomagnetic anomalies.) Strasbourg, Thèse, 1921.

M. R. Guillen. Sur l'existence du dimère $O_4$ dans l'oxygène liquide. *C.R.* 198, 1486, 1934.

†W. H. de Haas, E. C. Wiersma and W. H. Capel. The Determination of the susceptibility of Erbium Sulphate at Low Temperatures. *Leiden Comm.* 201c.

W. Heisenberg. Mehrkörperproblem und Resonanz in der Quantenmechanik. *Zs. f. Phys.* 38, 411, 1926.

Honda and Ishiwara. (Susceptibilities of paramagnetic salts.) *Sc. Rep. Tohôku Imp. Univ.* 4, 215, 1915.

Ishiwara. (Susceptibilities of paramagnetic salts.) *Sc. Rep. Tohôku Imp. Univ.* 3, 303, 1914.

L. C. Jackson. (Susceptibility of ammonium manganese sulphate.) *Phil. Trans.* 224, 1, 1923.

H. Kamerlingh Onnes and E. Oosterhuis. On the Susceptibility of Gaseous Oxygen at Low Temperatures. *Leiden Comm.* 134c, 1913.

—— —— On Paramagnetism at Low Temperatures. *Leiden Comm.* 139c, 1914.

—— —— On Paramagnetism at Low Temperatures. *Leiden Comm.* 129b, 1914.

†H. Kamerlingh Onnes and A. Perrier. Apparatus for the General Cryomagnetic investigation of Substances of Small Susceptibility. *Leiden Comm.* 139a, 1914.

—— —— The Susceptibility of Liquid Mixtures of Oxygen and Nitrogen. *Leiden Comm.* 139d, 1914.

—— —— On Para- and Diamagnetism at very Low Temperatures. *Leiden Comm.* 122a, 1911.

L. Landau. Eine mögliche Erklärung der Feldabhängigkeit der Suszeptibilität bei niedrigen Temperaturen. *Sow. Phys.* 4, 675, 1933.

G. N. Lewis. (Polymerisation of $O_2$.) *Journ. Amer. Chem. Soc.* 46, 2027, 1924.

K. F. Niessen. On the Saturation of the Electric and Magnetic Polarisation of Gases in Quantum Mechanics. *Phys. Rev.* 34, 253, 1929.

†E. Oosterhuis. Modification in the Cryogenic Apparatus of Kamerlingh Onnes and Perrier. *Leiden Comm.* 139b, 1914.

# BIBLIOGRAPHY

A. PRIKHOTKO, M. RUHEMANN and A. FEDERITENKO. The Absorption Spectrum of Solid Oxygen. I. *Sow. Phys.* **7**, 410, 1935.

H. SALOW and W. STEINER. Absorption Spectrum of Oxygen at High Pressure and the existence of $O_4$-molecules. *Nature*, **134**, 463, 1934.

†F. SIMON and J. AHARONI. Magnetische Untersuchungen an sorbierten Gasen. *Zs. f. phys. Chem.* (B), **4**, 175, 1909.

THÉODORIDÈS. (Susceptibility of $CoSO_4$, $Fe(SO_4)3/2$, $MnSO_4$, $CoCl_2$, $NiCl_2$.) *Journ. de Phys.* **3**, 1, 1922.

O. N. TRAPEZNIKOWA and L. W. SCHUBNIKOW. Ueber die Anomalie der spezifischen Wärme von wasserfreiem Eisenchlorid. *Sow. Phys.* **7**, 66, 1935.

—— —— Ueber die Anomalie der spezifischen Wärme von wasserfreiem $CrCl_3$. *Sow. Phys.* **7**, 255, 1935.

—— —— Anomaly in the Specific Heat of Ferrous Chloride at the Curie Point. *Nature*, **134**, 378, 1934.

J. H. VAN VLECK. Magnetic Susceptibilities in Quantum Mechanics. *Phys. Rev.* **31**, 587, 1928.

E. C. WIERSMA and C. J. GORTER. Remarks on the Susceptibility of Oxygen Gas. *Physica*, **12**, 316, 1932.

†E. C. WIERSMA, W. J. DE HAAS and W. H. CAPEL. On the Change of the Magnetic Moment of NO with the Temperature. *Leiden Comm.* 212b, 1931.

—— —— —— On the Magnetic Susceptibility of Oxygen at Low Pressure. *Leiden Comm.* 215b, 1931.

†E. C. WIERSMA and H. R. WOLTJER. A Horizontal Cryostat for the Measurement of Magnetic Susceptibilities at Low Temperatures. *Leiden Comm.* 201c, 1929.

†H. R. WOLTJER. On the Determination of Magnetisation at very Low Temperatures. *Leiden Comm.* 167b.

H. R. WOLTJER and H. KAMERLINGH ONNES. (Susceptibility of $CrCl_3$ etc.) *Leiden Comm.* 173b, c.

H. R. WOLTJER, C. W. COPPOOLSE and E. C. WIERSMA. On the Magnetic Susceptibility of Oxygen as Function of Temperature and Density. *Leiden Comm.* 201d, 1929.

H. R. WOLTJER and H. KAMERLINGH ONNES. On the Magnetisation of Gadolinium Sulphate at Temperatures obtainable with Liquid Helium. *Leiden Comm.* 167c, 1923.

H. R. WOLTJER and E. C. WIERSMA. On Anomalous Magnetic Properties at Low Temperatures. Anhydrous Ferrous Chloride. *Leiden Comm.* 201a, 1929.

## CHAPTER III

P. DEBYE. Einige Bemerkungen zur Magnetisierung bei tiefen Temperaturen. *Ann. der Physik* (4), **81**, 1154, 1926.

—— Die magnetische Methode zur Erzeugung tiefster Temperaturen. *Phys. Zs.* **35**, 923, 1934.

W. J. DE HAAS. Das Erreichen niedriger Temperaturen mittels adiabatischer Entmagnetisierung. *Naturw.* **21**, 732, 1933; *Nature*, **132**, 372, 1933.

BIBLIOGRAPHY

W. J. DE HAAS. Ueber extrem niedrige Temperaturen. *Naturw.* 21, 732, 1933.

W. J. DE HAAS and E. C. WIERSMA. Adiabatic Demagnetisation of some Paramagnetic Salts. *Physica*, 2, 335, 1935.

W. J. DE HAAS, E. C. WIERSMA and H. A. KRAMERS. Experiments on Adiabatic Cooling of Paramagnetic Salts in Magnetic Fields. *Physica*, 1, 1, 1933.

W. F. GIAUQUE. A Proposed Method of Producing Temperatures considerably below 1° abs. *Journ. Amer. Chem. Soc.* 49, 1864, 1927.

—— Paramagnetism and the Third Law of Thermodynamics. *Journ. Amer. Chem. Soc.* 49, 1870, 1927.

W. G. GIAUQUE and D. P. MACDOUGAL. Attainment of Temperatures below 1° abs. by Demagnetisation of $Gd_2(SO_4)_2.8H_2O$. *Phys. Rev.* 43, 768, 1933.

—— —— The Production of Temperatures below 1° abs. by Adiabatic Demagnetisation of Gadolinium Sulphate. *Journ. Amer. Chem. Soc.* 57, 1175, 1935.

W. H. KEESOM. The Thermodynamic Temperature Scale below 1° K. *Physica*, 2, 805, 1935; *Journ. de Phys.* 5, 373, 1934.

N. KÜRTI. Ueber das thermische und magnetische Verhalten des Gadoliniumsulfats im Temperaturgebiet des flüssigen Heliums. *Zs. f. phys. Chem.* (B), 20, 305, 1933.

N. KÜRTI and F. SIMON. Kalorimetrischer Nachweis einer Termaufspaltung im Gadoliniumsulfat. *Naturw.* 21, 178, 1933.

—— —— A Simple Arrangement for the Magnetic Cooling Method. *Physica*, 1, 1107, 1934.

—— —— Supraconductivity of Cadmium. *Nature*, 133, 907, 1934.

†—— —— Experiments at very Low Temperatures obtained by the Magnetic Method. I. The Production of the Low Temperatures. *Proc. Roy. Soc.* (A), 149, 152, 1935.

—— —— Further Experiments with the Magnetic Cooling Method. *Nature*, 135, 31, 1935.

—— —— Experiments at very Low Temperatures obtained by the Magnetic Method. II. New Supraconductors. *Proc. Roy. Soc.* (A), 151, 610, 1935.

PART IV

CHAPTER I

*E. GRÜNEISEN. Metallische Leitfähigkeit. *Handbuch der Physik*, 13, 1, 1927.

*W. HUME ROTHERY. The Metallic State. Oxford, Clarendon Press, 1931.

F. BLOCH. Ueber die Quantenmechanik der Elektronen in Kristallgittern. *Zs. f. Phys.* 52, 555, 1928.

—— Zum elektrischen Widerstandsgesetz bei tiefen Temperaturen. *Zs. f. Phys.* 59, 208, 1928.

# BIBLIOGRAPHY

†H. BREMMER. Onderzoekingen over de Warmtegeleiding bij lage temperaturen. Leiden Thesis, 1934.

P. EHRENFEST. Bemerkung über den Diamagnetismus von festem Wismut. *Zs. f. Phys.* **58**, 719, 1929.

E. GRÜNEISEN and H. REDDEMANN. Elektronen- und Gitterleitung beim Wärmefluss in Metallen. *Phys. Zs.* **35**, 959, 1934.

W. J. DE HAAS and VAN ALPHEN. The Dependence of the Susceptibility of Diamagnetic Metals upon the Field. *Leiden Comm.* 212a, 1930.

—— —— The Dependence of the Susceptibility of Bismuth Single Crystals upon the Field. *Leiden Comm.* 220d, 1932.

W. J. DE HAAS and BIERMASZ. Thermal Conductivity of Quartz at Low Temperatures. *Physica,* **2**, 673, 1935.

W. J. DE HAAS and J. W. BLOM. On the Change of Resistance of Single Crystals of Gallium in a Magnetic Field. I and II. *Leiden Comm.* 229b, 231b; *Physica,* **1**, 465, 1934.

W. J. DE HAAS and J. DE BOER. The Electrical Resistance of Platinum at Low Temperatures. *Leiden Comm.* 213c; *Physica,* **1**, 609, 1934.

W. J. DE HAAS and H. BREMMER. Thermal Conductivity of Lead and Tin at Low Temperatures. *Leiden Comm.* 214d, 1931.

—— —— The Thermal Conductivity of Indium. *Leiden Comm.* 220b, 1932.

—— —— The Conduction of Heat of Lead-Thallium at Low Temperatures. *Leiden Comm.* 220c, 1932.

†W. J. DE HAAS and W. H. CAPEL. A Method for the Determination of the Thermal Resistance of Metal Single Crystals at Low Temperatures. *Physica,* **1**, 725, 1934.

W. J. DE HAAS and VOOGD. Rapport No. 10, Congrès international de l'Électricité, Paris, 1932. *Leiden Comm.* Suppl. 73b, 1933.

W. J. DE HAAS, S. AOYAMA and H. BREMMER. Thermal Conductivity of Tin at Low Temperatures. *Leiden Comm.* 214a, 1931.

H. KAMERLINGH ONNES and W. TUYN. Data concerning the Electrical Resistance of Elementary Substances at Temperatures below −80° C. *Leiden Comm.* Suppl. 58, 1926.

P. KAPITZA. Specific Resistance of Bismuth Crystals and its Change in Strong Magnetic Fields. *Proc. Roy. Soc.* (A), **119**, 358, 1928.

M. KOHLER. Magnetische Widerstandsänderung in Metallkristallen. *Zs. f. Phys.* **95**, 365, 1935.

W. MEISSNER and B. VOIGT. Widerstand der reinen Metalle bei tiefen Temperaturen. *Ann. der Phys.* (5), **7**, 761, 892, 1930.

R. PEIERLS. Zur Theorie der elektrischen und thermischen Leitfähigkeit von Metallen'. *Ann. der Phys.* (5), **4**, 121, 1930.

—— Das Verhalten metallischer Leiter in starken Magnetfeldern. *Leipziger Vorträge,* p. 75, 1930.

—— Zur Frage des elektrischen Widerstandsgesetzes bei tiefen Temperaturen. *Ann. der Phys.* (5), **12**, 154, 1932.

L. SCHUBNIKOW and W. J. DE HAAS. A New Phenomenon in the Change of Resistance in a Magnetic Field of Single Crystals of Bismuth. *Nature,* Oct. 4th, 1930.

BIBLIOGRAPHY

L. Schubnikow and W. J. de Haas. Magnetische Widerstandsvergrösse-
rung in Einkristallen von Wismuth bei tiefen Temperaturen. *Leiden
Comm.* 207 a, 1930.
—— —— Neue Erscheinungen bei der Widerstandsänderung von Wis-
mutheinkristallen im Magnetfeld bei der Temperatur von flüssigem
Wasserstoff. I und II. *Leiden Comm.* 207 d, 1930; 210 a, 1930.
—— —— Die Widerstandsänderung von Wismutheinkristallen im
Magnetfeld bei der Temperatur des flüssigen Stickstoffs. *Leiden Comm.*
210 b, 1930.
A. Sommerfeld. Zur Elektronentheorie der Metalle auf Grund der
Fermi'schen Statistik. *Zs. f. Phys.* 47, 1, 1928.
O. Stierstadt. Die acht elektrischen Hauptleitfähigkeiten des Bi-
Kristalls im Magnetfeld. *Zs. f. Phys.* 95, 335, 1935.

CHAPTER II

*W. Meissner. Supraleitfähigkeit. *Erg. der exakt. Naturw.* xi, 219,
1932.
J. F. Allen. The Supraconductivity of Alloy Systems. *Phil. Mag.* 16,
1005, 1933.
E. F. Burton. Magnetic Properties of Supraconductors. *Nature*, 133, 684,
1934.
C. Gorter. Some Remarks on the Thermodynamics of Supraconductivity.
*Arch. Musée Teyler,* 7, 378, 1933.
C. Gorter and H. Casimir. Zur Thermodynamik des supraleitenden
Zustands. *Physica,* 1, 306, 1934; *Phys. Zs.* 35, 963, 1934.
W. J. de Haas and J. M. Casimir-Jonker. Untersuchungen über den
Verlauf des Eindringens eines transversalen Magnetfeldes in einen
Supraleiter. *Leiden Comm.* 229 d; *Physica,* 1, 291, 1934.
W. J. de Haas and J. Voogd. On the Steepness of the Transition Curve
of Supraconductors. *Leiden Comm.* 214 c, 1931.
W. J. de Haas, E. van Aubel and J. Voogd. A Supraconductor Con-
sisting of Two Non-Supraconductors. *Leiden Comm.* 197 c, 1929.
W. J. de Haas, J. Voogd and J. M. Jonker. Quantitative Untersuchung
über einen möglichen Einfluss der Achsenorientierung auf die mag-
netische Uebergangsfigur. *Leiden Comm.* 229 c; *Physica,* 1, 281,
1934.
T. C. Keeley, K. Mendelssohn and J. R. Moore. Experiments on
Supraconductors. *Nature,* 134, 773, 1934.
W. H. Keesom and J. N. van den Ende. The Specific Heats of Lead and
Bismuth. *Leiden Comm.* 203 d, 1929; 213 c, 1931.
—— —— The Atomic Heats of Tin and Zinc. *Leiden Comm.* 219 c, 1931.
W. H. Keesom and J. A. Kok. On the Change in the Specific Heat of Tin
when becoming Supraconductive. *Leiden Comm.* 221 e, 1932.
—— —— Measurements of the Specific Heat of Thallium at Liquid
Helium Temperatures. *Leiden Comm.* 230 c; *Physica,* 1, 175, 1933.
—— —— Further Calorimetric Experiments on Thallium. *Leiden Comm.*
232 a; *Physica,* 1, 595, 1934.

## BIBLIOGRAPHY

W. H. KEESOM and J. A. KOK. Measurements of the Latent Heat of Thallium connected with the Transition, in a Constant External Magnetic Field, from the Supraconductive to the Non-supraconductive State. *Leiden Comm.* 230e; *Physica*, **1**, 503, 1934.

F. and H. LONDON. The Electromagnetic Equations of the Supraconductor. *Proc. Roy. Soc.* (A), **149**, 71, 1935.

—— Supraleitung und Diamagnetismus. *Physica*, **2**, 341, 1935.

J. C. MACLENNAN. Electric Supraconduction in Metals. *Nature*, Dec. 10th, 1932.

W. MEISSNER. Bericht über neuere Arbeiten zur Supraleitfähigkeit. *Phys. Zs.* **35**, 931, 1934.

W. MEISSNER and R. OCHSENFELD. Neuer Effekt bei Eintritt der Supraleitfähigkeit. *Naturw.* **21**, 787, 1933.

K. MENDELSSOHN and J. D. BABBITT. Persistent Currents in Supraconductors. *Nature*, **133**, 459, 1934.

—— —— Magnetic Behaviour of Supraconducting Tin Spheres. *Proc. Roy. Soc.* (A), **151**, 316, 1935.

K. MENDELSSOHN and J. R. MOORE. Magneto-Caloric Effect in Supraconducting Tin. *Nature*, **133**, 413, 1934.

—— —— Specific Heat of a Supraconducting Alloy. *Proc. Roy. Soc.* (A), **151**, 334, 1935.

K. MENDELSSOHN and F. SIMON. Ueber den Energieinhalt des Bleis in der Nähe des Sprungpunktes der Supraleitfähigkeit. *Zs. f. phys. Chem.* (B), **16**, 72, 1932.

J. N. RJABININ and L. W. SCHUBNIKOW. Verhalten eines Supraleiters im magnetischen Feld. *Sow. Phys.* **5**, 641, 1934.

—— —— Ueber die Abhängigkeit der magnetischen Induktion des supraleitenden Bleis vom Feld. *Sow. Phys.* **6**, 557, 1934.

—— —— Dependence of Magnetic Induction on the Magnetic Field in Supraconducting Lead. *Nature*, **134**, 286, 1934.

—— —— Magnetic Properties and Critical Currents of Supraconducting Alloys. *Sow. Phys.* **7**, 122, 1935.

—— —— Magnetic Induction in a Supraconducting Lead Crystal. *Nature*, **135**, 109, 1935.

L. W. SCHUBNIKOW and W. J. CHOTKEWITSCH. Spezifische Wärme von supraleitenden Legierungen. *Sow. Phys.* **6**, 605, 1934.

F. G. A. TARR and J. O. WILHELM. Effective Permeability of Supraconductors. *Trans. Roy. Soc. Canada*, **3**, 61, 1934.

# ADDENDA

## Literature published between the end of 1935
and May 1937.

### PART I

P. BOURBO and I. ISCHKIN. Untersuchungen über das Gleichgewicht von Flüssigkeit und Dampf des Systems Argon-Sauerstoff. *Sow. Phys.* **10**, 271, 1936; *Physica*, **3**, 1067, 1936.

V. G. FASTOVSKY and L. A. GIRSKAYA. Adsorption of neon and helium. *Journ. Chem. Ind.* (Russian), **14**, 358, 1937.

A. FEDORITENKO and M. RUHEMANN. Equilibrium diagrams of helium-nitrogen mixtures. *Techn. Phys. of U.S.S.R.* **4**, 36, 1937.

S. V. GERSH. Low Temperature Engineering. *Moscow O.N.T.I.* Part I, 1936; Part II, 1937.

F. HENNING and J. OTTO. Dampfdruckkurven und Tripelpunkte im Temperaturgebiet von 14° bis 90° abs. *Phys. Zs.* **37**. 633, 1936.

—— —— Das Platinwiderstandsthermometer als sekundäres Temperaturnormal zwischen 14° und 90° abs. *Phys. Zs.* **37**, 639, 1936.

W. H. KEESOM and A. P. KEESOM. Measurements concerning the specific heat of solid helium and the melting heat of helium. *Physica*, **3**, 105, 1936.

W. H. KEESOM and A. BIJL. Determination of the vapour pressure of liquid nitrogen below 1 atm. and of solid nitrogen $\beta$. The boiling point and triple point of nitrogen. *Physica*, **4**, 305, 1937.

J. K. KRITSCHEWSKY and N. S. TOROTSCHESCHNIKOW. Thermodynamik des Flüssigkeitdampf-Gleichgewichts im Stickstoff-Sauerstoffsystem. *Zs. f. phys. Chem.* (A), **176**, 338, 1936.

V. I. ROMANOW and V. G. FASTOVSKY. Obtaining neon-helium mixture. *Journ. Chem. Ind.* (Russian), **14**, 105, 1937.

M. RUHEMANN and A. FEDORITENKO. Use of $i$-$x$-diagrams in separating helium and nitrogen. *Journ. Chem. Ind.* (Russian), **14**, 28, 1937.

H. VEITH and E. SCHRÖDER. Schmelzdiagramme einiger binärer Systeme aus kondensierten Gasen. *Zs. f. phys. Chem.* (A), **179**, 16, 1937.

### PART II

C. T. ANDERSON. Heat capacities of vanadium and its oxides. *Journ. Amer. Chem. Soc.* **58**, 565, 1936.

—— The heat capacities of molybdenite and pyrite at low temperatures. *Journ. Amer. Chem. Soc.* **59**, 486, 1937.

—— The heat capacities of Cr and some of its compounds at low temperatures. *Journ. Amer. Chem. Soc.* **59**, 488, 1937.

ADDENDA

A. BIJL. Properties of the condensed phases of helium and hydrogen. *Physica*, **4**, 329, 1937.

M. BLACKMAN. On the vibrational spectrum of a three-dimensional lattice. *Proc. Roy. Soc.* (A), **159**, 416, 1937.

O. L. BROWN and G. G. MANOV. The heat capacity and entropy of carbon disulfide. *Journ. Amer. Chem. Soc.* **59**, 500, 1937.

E. F. BURTON and W. F. OLIVER. The crystal structure of ice at low temperatures. *Proc. Roy. Soc.* (A), **153**, 166, 1936.

E. D. EASTMAN and W. C. MCGAVOCK. The heat capacity and entropy of rhombic and monoclinic sulphur. *Journ. Amer. Chem. Soc.* **59**, 145, 1937.

W. F. GIAUQUE and R. C. ARCHIBALD. The entropy of water, the heat capacities of $Mg(OH)_2$ and MgO and the heat of water formation. *Journ. Amer. Chem. Soc.* **59**, 561, 1937.

W. F. GIAUQUE and R. W. BLUE. Heat capacity of solid and liquid $H_2S$. *Journ. Amer. Chem. Soc.* **58**, 831, 1936.

W. F. GIAUQUE and C. J. EGAN. Carbon dioxide; the heat capacity of the solid and other thermodynamic properties. *Chem. Phys.* **4**, 45, 1936.

W. F. GIAUQUE and J. W. STOUT. Heat capacity and entropy of ice. *Journ. Amer. Chem. Soc.* **58**, 1144, 1936.

W. H. KEESOM and A. P. KEESOM. Specific heat of solid helium and the melting heat of helium. *Physica*, **3**, 105, 1936.

—— —— Heat conductivity of liquid helium. *Physica*, **3**, 359, 1936.

W. H. KEESOM and K. W. TACONIS. Debye Scherrer exposures of liquid helium. *Physica*, **4**, 28, 256, 1937.

—— —— The structure of solid $\gamma$-$O_2$. *Physica*, **3**, 141, 1936.

—— —— The crystal structure of $Cl_2$. *Physica*, **3**, 237, 1936.

J. D. KEMP and W. F. GIAUQUE. The heat capacity of carbon disulphide and the third law of thermodynamics. *Journ. Amer. Chem. Soc.* **59**, 79, 1937.

J. D. KEMP and K. S. PITZER. The entropy of ethane and the third law of thermodynamics. *Journ. Amer. Chem. Soc.* **59**, 276, 1937.

J. A. KOK and W. H. KEESOM. The atomic heats of Pt and Cu from $1\cdot2°$ to 20° K. *Physica*, **3**, 1035, 1936.

L. LANDAU. Theorie der Phasenumwandlungen. I und II. *Sow. Phys.* **11**, 26, 545, 1937.

C. H. LANDER and J. V. HOWARD. The tensile properties of solid Hg. *Proc. Roy. Soc.* (A), **156**, 411.

F. LONDON. On condensed helium at absolute zero. *Proc. Roy. Soc.* (A), **153**, 576, 1936.

E. A. LONG and J. D. KEMP. Heat capacity and entropy of $D_2O$. *Journ. Amer. Chem. Soc.* **58**, 1829, 1937.

R. L. MCFARLAN. The crystal structure of ice. II and III. *Chem. Phys.* **4**, 60, 253, 1936.

S. A. MCNEIGHT and C. P. SMITH. Molecular rotation in solid arsine and other hydrides. *Journ. Amer. Chem. Soc.* **58**, 1723, 1936.

N. F. MOTT. The resistance and thermoelectric properties of the transition metals. *Proc. Roy. Soc.* (A), **156**, 368, 1936.

⟨ 310 ⟩

ADDENDA

R. OVERSTREET and W. F. GIAUQUE. The heat capacity and entropy of ammonia. *Journ. Amer. Chem. Soc.* **59**, 254, 1937.

G. PANKOW. Anomalie der spezifischen Wärme bei Li. *Helv. Phys. Acta*, **9**, 87, 1936.

L. PAULING. The structure and entropy of ice and other crystals with some randomness of atomic arrangement. *Journ. Amer. Chem. Soc.* **57**, 2680, 1935.

L. W. SCHUBNIKOW and A. A. KIKOIN. Optische Untersuchungen an flüssigem Helium. II. *Sow. Phys.* **10**, 119, 1936.

A. SMITS and G. J. MÜLLER. The low temperature transformation of heavy ammonium chloride. *Nature*, **139**, 804, 1937.

C. P. SMYTH and S. A. MCNEIGHT. Non-rotation of molecules in a number of solids. *Journ. Amer. Chem. Soc.* **58**, 1718, 1936.

—— —— Molecular rotation in solid aliphatic alcohols. *Journ. Amer. Chem. Soc.* **58**, 1597, 1936.

C. C. STEPHENSON and W. F. GIAUQUE. Phosphine. Test of the third law of thermodynamics. *Chem. Phys.* **5**, 149, 1937.

E. C. STONER. Collective electron specific heat and spin paramagnetism in metals. *Proc. Roy. Soc.* (A), **154**, 656, 1936.

O. TRAPEZNIKOWA, L. SCHUBNIKOW and G. MELJUTIN. Ueber die Anomalie der spezifischen Wärmen von wasserfreiem $CrCl_3$. *Sow. Phys.* **9**, 237, 1936.

O. TRAPEZNIKOWA and G. MELJUTIN. Die spezifische Wärme von wasserfreiem $MnCl_2$. *Sow. Phys.* **11**, 55, 1937.

L. VEGARD. The structure of solid oxygen. *Nature*, **136**, 720, 1935.

G. WEIGLÉ and H. SAINI. La structure du bromure d'ammonium à basses températures. *Helv. Phys. Acta*, **9**, 514, 1936.

R. K. WITT and J. D. KEMP. The heat capacity of ethane from 15° K. to boiling point. *Journ. Amer. Chem. Soc.* **59**, 276, 1937.

W. A. YAGER and S. O. MORGAN. Transitions in camphor and chemically related compounds. Dipole rotation in crystalline compounds. *Journ. Amer. Chem. Soc.* **57**, 2071, 1935.


PART III

J. BECQUEREL, W. J. DE HAAS and J. VAN DE HANDEL. Pouvoir rotatoire paramagnétique de l'éthylsulfate de dysprosium hydraté et saturation paramagnétique. *Physica*, **3**, 1133, 1936.

—— —— —— Pouvoir rotatoire paramagnétique de l'éthylsulfate d'erbium hydraté et saturation paramagnétique. *Physica*, **4**, 345, 1937.

C. W. CLARK and W. H. KEESOM. The heat capacity of gadolinium sulphate from 1·0° to 20·5° K. *Physica*, **2**, 1075, 1935.

W. F. GIAUQUE and D. P. MACDOUGAL. The thermodynamic temperature scale in the region below 1° absolute. *Phys. Rev.* **47**, 885, 1935.

W. J. DE HAAS and E. C. WIERSMA. On the determination of the thermodynamical temperature scale below 1° K. *Physica*, **3**, 491, 1936.

⟨ 311 ⟩

ADDENDA

N. Kürti, B. V. Rollin and F. Simon. Preliminary experiments on temperature equilibrium at very low temperatures. *Physica*, 3, 266, 1936.

N. Kürti, P. Lainé, B. Rollin and F. Simon. Sur l'apparition de ferromagnétisme dans quelques sels paramagnétiques à de très basses températures. *C.R.* May 11th, 1936.

B. G. Lasarew and L. W. Schubnikow. Das magnetische Moment des Protons. *Sow. Phys.* 11, 445, 1937.

L. W. Schubnikow and S. S. Schalyt. Ferromagnetische Eigenschaften einiger paramagnetischer Salze. *Sow. Phys.* 11, 566, 1937.

F. Simon. Application of low temperature calorimetry to radioactive measurements. *Nature*, 135, 763, 1935.


PART IV

H. Bremmer and W. J. de Haas. On the conduction of heat by some metals at low temperatures. *Physica*, 3, 672, 1936.

—— —— On the heat-conductivity of supraconducting alloys. *Physica*, 3, 692, 1936.

W. J. de Haas and H. Bremmer. Determination of the heat resistance of mercury at temperatures attainable with liquid helium. *Physica*, 3, 687, 1936.

W. J. de Haas, A. N. Gerritson and W. H. Capel. The thermal resistance of bismuth single crystals at low temperatures and in a magnetic field. *Physica*, 3, 1143, 1936.

W. J. de Haas and A. D. Engelkes. On the disturbance of the superconductive state by a magnetic field. Supplementary measurements. *Physica*, 4, 325, 1937.

W. J. de Haas and O. A. Guineau. On the transition of a monocrystalline tin sphere from the superconductive to the non-superconductive state. *Physica*, 3, 182, 534, 1936.

E. Justi and H. Scheffers. Ueber den elektrischen Widerstand des Goldes bei tiefen Temperaturen im magnetischen Transversalfeld. *Phys. Zs.* 37, 383, 475, 1936.

—— —— Die elektrische Anisotropie von Wolframeinkristallen bei tiefen Temperaturen im magnetischen Transversalfeld. *Phys. Zs.* 37, 700, 1936.

T. C. Keeley and K. Mendelssohn. Magnetic properties of supraconductors. *Proc. Roy. Soc.* (A), 154, 378, 1936.

W. H. Keesom and P. H. van Laer. Measurements of the latent heat of tin while passing from the superconductive to the non-superconductive state. *Physica*, 3, 371, 1936; 4, 487, 1937.

—— —— Relaxation phenomena in superconductivity. *Physica*, 4, 499, 1937.

L. Landau. Zur Theorie der Supraleitfähigkeit. *Sow. Phys.* 11, 129, 1937.

K. C. Mann and J. O. Wilhelm. The influence of magnetic fields on persistent currents in superconductors. *Trans. Roy. Soc. Canada*, 31, 19, 1937.

ADDENDA

K. Mendelssohn and J. G. Daunt. Supraconductivity of lanthanum. *Nature*, 139, 473, 1937.

K. Mendelssohn. The transition between the supraconductive and the normal state. I. Magnetic induction in mercury. *Proc. Roy. Soc.* (A), 155, 558, 1936.

K. Mendelssohn and R. B. Pontius. Note on magnetic hysteresis and time effects in supraconductors. *Physica*, 3, 327, 1936.

R. Peierls. Magnetic transition curves of supraconductors. *Proc. Roy. Soc.* (A), 155, 613, 1936.

L. Shubnikov. Destruction of supraconductivity by electric current and magnetic field. *Nature*, 138, 545, 1936.

L. W. Schubnikow and N. E. Alexejewski. Transition curve for the destruction of supraconductivity by an electric current. *Nature*, 138, 804, 1936.

L. W. Schubnikow, W. I. Chotkewitsch, J. D. Schepelew and J. N. Rjabinin. Magnetische Eigenschaften supraleitender Metallegierungen. *Sow. Phys.* 10, 165, 1936.

L. W. Schubnikow and W. I. Chotkewitsch. Kritische Werte des Feldes und des Stromes für die Supraleitfähigkeit des Zinns. *Sow. Phys.* 10, 231, 1936.

H. Grayson Smith. Resistance of a superconductor in the intermediate state. *Trans. Roy. Soc.* (A), 31, 31, 1937.

H. G. Smith and J. O. Wilhelm. Destruction of magnetic field around simply and multiply connected supraconductors. *Proc. Roy. Soc.* (A), 157, 132, 1936.

Printed in the United States
By Bookmasters